QUANTUM SPACE

JIM BAGGOTT is a freelance science writer. He was a lecturer in chemistry at the University of Reading but left to work with Shell International Petroleum Company and then as an independent business consultant and trainer. His many books include *Quantum Reality: The Quest for the Real Meaning of Quantum Mechanics—A Game of Theories* (OUP, 2020), *Quantum Space: Loop Quantum Gravity and the Search for the Structure of Space, Time, and the Universe* (OUP, 2018), *Mass: The Quest to Understand Matter from Greek Atoms to Quantum Fields* (OUP, 2017), *Origins: The Scientific Story of Creation* (OUP, 2015), *Higgs: The Invention and Discovery of the 'God Particle'* (OUP, 2012), *The Quantum Story: A History in 40 Moments* (OUP, 2011), and *A Beginner's Guide to Reality* (Penguin, 2005).

As any theoretical physicist will tell you, quantum gravity is the biggest scientific problem of our age.

The 20th century gave us two extraordinarily successful theories of physics: quantum mechanics, which works astonishingly well at very small scales, and general relativity, which beautifully explains gravity and the large-scale cosmos. The trouble is they don't work *together*. Quantum mechanics assumes an arena of space and time, as if everything were being played out against an invisible backdrop. General relativity dispenses with the backdrop: space and time are relative, and gravity is just the effect of matter moving in curved spacetime. Theorists in the 21st century seek to transcend both, by developing a single, joined-up account that can explain the behaviour of the universe at quantum scales. They seek a theory of quantum gravity.

There are two major approaches to quantum gravity, seen as rivals, though at root they have much in common. One, string theory, has been widely popularized, and arises primarily from the viewpoint of particle physics. This book is about the other approach, loop quantum gravity, or LQG. Less well known, it is gaining increasing interest and influence. Starting from general relativity, it borrows many ideas and techniques from particle theories, and predicts that space itself is quantum in nature. Time emerges as an evolving sequence of jumps in the geometry of quantum space which form a 'spinfoam'. It's all very abstract, but *Quantum Space* offers an opportunity to glimpse, without any mathematical technicalities, the deepest, most fundamental contemporary ideas concerning space, time, and the universe.

Lee Smolin and Carlo Rovelli have been leading players in the development of LQG, and Jim Baggott frames the story around the work and life experiences of these two famous physicists, who are close friends: their hopes, their frustrations, and their moments of triumph. This is also their story.

QUANTUM SPACE

LOOP QUANTUM GRAVITY AND THE **SEARCH** FOR THE **STRUCTURE** OF **SPACE, TIME,** AND THE **UNIVERSE**

JIM BAGGOTT

OXFORD
UNIVERSITY PRESS

Great Clarendon Street, Oxford, OX2 6DP,
United Kingdom

Oxford University Press is a department of the University of Oxford.
It furthers the University's objective of excellence in research, scholarship,
and education by publishing worldwide. Oxford is a registered trade mark of
Oxford University Press in the UK and in certain other countries

First published 2018
First published in paperback 2022

Impression: 1

Published in the United States of America by Oxford University Press
198 Madison Avenue, New York, NY 10016, United States of America

British Library Cataloguing in Publication Data

Data available

Library of Congress Cataloging in Publication Data

Data available

ISBN 978–0–19–880911–1 (Hbk.)
ISBN 978–0–19–880912–8 (Pbk.)

Printed and bound in Great Britain by
Clays Ltd, Elcograf S.p.A.

To Carlo and Lee
For entrusting me with your stories

CONTENTS

PART III: ELABORATIONS

PREFACE

Let's get one thing straight.

This is a book about loop quantum gravity, one of several contemporary approaches to the development of a quantum theory of gravity, perched right on the very edge of our current understanding of space, time, and the physical universe. One hopes that science at the frontiers will always make for entertaining reading but, make no mistake, like all such theories, as of today there is *not one single piece of observational or experimental evidence to support it.*[1]

You might then wonder why I think you ought to be interested in this.

Here's why. There's little doubting that in these first few decades of the twenty-first century we face some tremendous economic, political, and environmental challenges, some much more stubborn and intractable than others. But when it comes to our ability to comprehend the nature of space and time, to understand the very fabric of physical reality, *the quantum theory of gravity is simply the greatest scientific problem of our age.*[2] It addresses the ultimate 'big question' of existence. Resolving this problem demands a real depth of scientific expertise; it demands unique moments of insight and inspiration; and it demands intellectual creativity likely to be unsurpassed in the entire history of physics.

The reason is simple. Today we are blessed with two extraordinarily successful theories. The first is Albert Einstein's general theory of relativity, which describes the large-scale behaviour of matter in a curved spacetime. It tells us how *gravity* works: matter tells spacetime how to curve, and curved spacetime tells matter how to move. This theory is the basis for the so-called standard model of Big Bang cosmology. We use it to describe the evolution of our universe from almost the very 'beginning', which on current evidence happened about 13.8 billion years ago. The discovery of gravitational waves at the LIGO observatory in the USA (and now Virgo, in Italy) is only the most recent of this theory's many triumphs.

The second is quantum mechanics. This theory describes the properties and behaviour of matter and radiation at its smallest scales; at the level of molecules, atoms, sub-atomic, and sub-nuclear particles. In the guise of quantum field theory it is the basis for the so-called standard model of particle physics, which builds up all the visible constituents of the universe (including stars, planets, and us) out of collections of quarks, electrons, and force-carrying particles such as photons. It tells us how the other three forces of nature work: electromagnetism, the strong force, and the weak interaction. The discovery of the Higgs boson at CERN in Geneva is only the most recent of this theory's many triumphs.

But, while they are both highly successful, grand intellectual achievements, these two standard models are also riddled with holes. There's an awful lot they can't explain, and they leave a lot of important questions unanswered. If anything, their successes have only served to make the universe appear more elusive and mysterious, if not downright bizarre. The more we have learned, the less we seem to understand.

The two theories are also fundamentally incompatible. In the classical mechanics of Isaac Newton, objects exist and things happen within a 'container' of absolute space and time which somehow sits in the background. If we could take everything out of Newton's universe we must suppose that the empty container would remain. General relativity gets rid of this container. In Einstein's universe space and time become relative, not absolute, and the theory is said to be 'background independent'. Spacetime is dynamic; it *emerges* as a result of physical interactions involving matter and energy.

Quantum mechanics, though exasperatingly bizarre yet unfailingly accurate in its predictions, is formulated in a different way. Interactions involving the elementary particles of matter and radiation are assumed to take place in precisely the kind of absolute spacetime container that general relativity eliminates. Quantum mechanics is background-dependent.

And there you have it. We have a classical (non-quantum) theory of spacetime which is background-independent. And we have a quantum theory of matter and radiation which is background-dependent. Our two most successful theories of physics are built on incompatible interpretations of space and time. They are woven on different kinds of fabric, one co-generated by the physics and the other pre-supposed and absolute.

We have two incompatible descriptions but, as far as we know (and certainly as far as we can prove), we've only ever had one universe. This is a problem because we also know that in the first few moments following its birth in the Big Bang, the universe would have existed at the quantum scale, at the mercy of a quantum mechanics. Now, the fact that we can't explain the origin and earliest moments of the universe might not trouble you overmuch, but the track-record of physics in the past hundred years or so has encouraged us to have greater expectations. What we *need* is a quantum theory of gravity.

So, do I have your attention yet?

The Chinese philosopher Laozi once said that a journey of a thousand miles begins with a single step. The first thing we can do is recognize that the only way to bring together quantum mechanics and general relativity is to find a new fabric, a new way of conceiving of space and time, one that is compatible with physics on any scale.

Charged with a newfound sense of purpose, we must now choose which road to take. Do we start with the pre-supposed, absolute spacetime fabric of quantum mechanics? Or do we start with the co-generated fabric of general relativity?

In the past forty years or so, judgments concerning the ease of passage along these two roads have split the theoretical physics community along essentially tribal lines. This split is very visible in a recent attempt to map the relationships between all the different ways of developing a quantum theory of gravity, which identified two distinct 'fundamental' branches: string theory and loop quantum gravity.[3] This divide isn't simply the result of differences of opinion between general relativists and particle theorists, as theorists on either side frequently borrow ideas and techniques from both general relativity and quantum field theory.

It is, however, true to say that the theoretical physics community is dominated by particle theorists, and particle theorists tend to favour the string theory approach. In the past twenty years or so, their highly successful PR has spilled into the popular science literature, with the result that few readers are even aware that there's more than one game in town, or more than one road that can be taken. For example, in one recent popular book about gravity, loop quantum gravity is mentioned only in passing, relegated to a footnote.[4] There are all sorts of reasons for this, and I will discuss some of these in what follows.

This book is about the road less travelled. It starts from general relativity, borrows ideas from quantum chromodynamics, and involves finding a way to turn the result into a quantum field theory of gravity. At the destination we find a fabric in which space is not continuous, but quantized. It comes in 'lumps' just like matter and radiation. The fabric is a system of interlinking 'loops' of gravitational force which form a 'spin network'. There are fundamental limits on the geometries of these loops, which define quanta of the area and volume of space in terms of something called the Planck length, which is about 1.6×10^{-35} metres, or about a hundredth of a billionth of a billionth of the diameter of a proton.

Different spin networks—different ways of interlinking the loops—define different quantum states of the geometry of space. The evolution of spin networks (the changing connections between one geometry and the next) then gives rise to a *spinfoam*. Adding spinfoams in something called a superposition describes an emergent spacetime, a fabric co-generated by the quantum physics.

This is loop quantum gravity, or LQG for short. It is now thirty years old and currently occupies the attentions of about thirty research groups around the world. The road from relativity has been difficult, with many highs and lows. There remain many challenges yet to be overcome, not least that of finding a way to torture the theory into providing one or more definitive empirical tests. But as Carlo Rovelli, one of the principal architects of LQG, explained a little while ago, 'the situation in quantum gravity is in my opinion...far better than twenty-five years ago, and, one day out of two, I am optimistic.'[5]

Readers of popular science may have heard about LQG from Lee Smolin, another of its principal architects, whose *Three Roads to Quantum Gravity* was published in 2000. Smolin briefly touched on LQG again in *The Trouble with Physics*, published ten years ago, and most recently in *Time Reborn*. Rovelli mentions LQG in his best-selling *Seven Brief Lessons on Physics*, and in his most recent book *Reality is Not What It Seems*.

My mission in *Quantum Space* is to correct an imbalance in public perception. I want to persuade you that LQG is not only a good game, it offers a genuine, credible alternative to the string theory approach. To do this I will share with you a little more detail about the theory than Smolin and Rovelli have so far shared in their own popular books. I not only want to give you some sense for what LQG tells us about space, time, and the universe, but also *how* and *why* it tells us these things.

In researching and writing this book I've been very fortunate to receive considerable encouragement, support, and insight from both Smolin and Rovelli. *This book is their story*, but we also need to get a couple of other things straight. LQG is the result of a collaboration involving many theorists over many years of effort. I've tried as far as possible to acknowledge as much of this effort as is feasible in a popular presentation, and can only offer my sincere apologies in advance to any member of the community reading this who feels that their efforts are under-represented or, even worse, overlooked. By the same token, as this book focuses principally on the efforts of two prominent contributors, it is not intended to provide a comprehensive summary of everything that's been done in the name of LQG.[6]

The book is structured in three parts. Part I sets the scene. It tells us about the things that Smolin and Rovelli learned about relativity, quantum mechanics, and Big Bang cosmology as young students and then as mature theorists. Readers already familiar with this background can safely skip it (but I hope they won't). Part II tells the story of the birth and evolution of LQG, starting with efforts to bring relativity and quantum mechanics together in the late 1950s, through Abhay Ashtekar's discovery of the 'new variables' that would make this possible, to the collaboration among Ashtekar, Smolin, and Rovelli (and many others) which yielded quanta of area and volume and the spinfoam formalism towards the turn of the previous century. Part III brings us reasonably up to date. It summarizes efforts to perform calculations of familiar physical quantities using LQG and the implications of the theory for quantum cosmology and the physics of black

holes. On this part of the journey we will also explore the interpretation of quantum mechanics and the reality (or otherwise)
of time.

I want to be straight with you about one final thing. Like the
string or M-theory framework, LQG is still a work in progress.
It is not finished and we don't yet have all the answers. Smolin
and Rovelli are, of course, enthusiasts, and although I've tried
to take a balanced view, a lot of their enthusiasm is inevitably
reflected in my choice of words. But it is important not to get
too carried away. Many other theorists who have been involved
in various stages of the journey have since lost faith, the optimism of the late 1990s giving way to more sober (and sombre)
assessments. Some have chosen to leave the field entirely and
work on different problems. I hope that readers will at least get
some sense of the scale of the challenge—chasing a theory of
quantum gravity is most definitely *not* for the faint of heart. The
book closes with a three-way exchange among Smolin, Rovelli,
and myself which looks back at recent history, and forward to
the future.

There's a lot at stake. The great revolutions in science that have
shaped the way we seek to comprehend reality have profoundly
changed the way we think about space, time, and the universe.
Could another revolution be close at hand?

This book would not have been possible had Lee and Carlo not
entrusted me with their stories. It's therefore a real pleasure to
acknowledge their commitment to this project, reading over my
shoulder as I worked on the manuscript, nudging me in the right
direction and putting me right when I got it wrong. Having said
that, it's important for you to know that the views expressed in
this book are entirely my own, and whilst Lee and Carlo agree with
much of what I've written, you shouldn't assume they agree with
everything.

In addition to thanking Lee and Carlo, I also need to acknowledge the efforts of many other busy scientists who gave up their valuable time to read through my draft manuscript, correct many of my misinterpretations and mistakes, and add insights of their own. These include Abhay Ashtekar at Pennsylvania State University, John Baez at the University of California, Riverside, Martin Bojowald at Pennsylvania State University, Alejandro Corichi at the National Autonomous University of Mexico, George Ellis at the University of Cape Town, Ted Jacobson at the University of Maryland, Kirill Krasnov at the University of Nottingham, Jorge Pullin at Louisiana State University, and Peter Woit at Columbia University.

Now, LQG is a theory that is far from complete. This means that even those who have been involved most closely in its development don't all agree on the answers to the theory's many open questions. In order to produce a hopefully coherent, readable narrative about a subject in which virtually everything can be challenged, I've had to be somewhat selective in what to present. I'm pretty sure I haven't got this right all the time, and it goes without saying that I'm happy to take the credit for all those errors that remain.

I must also once more acknowledge my debts to Latha Menon, my editor at Oxford University Press, and to Jenny Nugee, who have again worked industriously behind the scenes to produce the book you now hold in your hands. Without their efforts, the book would certainly have been poorer.

Shall we begin?

Jim Baggott
July 2018

LIST OF ABBREVIATIONS

ADM	Arnowitt, Deser, Misner
ATLAS	A Toroidal LHC Apparatus (detector)
CDM	cold dark matter
CERN	Conseil Européen pour la Recherche Nucléaire
CMS	Compact Muon Solenoid (detector)
COBE	Cosmic Background Explorer
CODATA	International Council for Science Committee on Data for Science and Technology
GeV	giga electron volt
GUT	grand unified theory
Λ-CDM	lambda-cold dark matter
LHC	large hadron collider
LQC	loop quantum cosmology
LQG	loop quantum gravity
MeV	mega electron volt
MSSM	minimum supersymmetric standard model
NSF	National Science Foundation
QCD	quantum chromodynamics
QED	quantum electrodynamics
SLAC	Stanford Linear Accelerator Center
SUSY	supersymmetry
TeV	tera electron volt
WMAP	Wilkinson Microwave Anisotropy Probe

ABOUT THE AUTHOR

Jim Baggott is an award-winning science writer. A former academic scientist, he now works as an independent business consultant but maintains a broad interest in science, philosophy, and history and continues to write on these subjects in his spare time. His previous books have been widely acclaimed and include the following:

Quantum Reality: The Quest for the Real Meaning of Quantum Mechanics—A Game of Theories, Oxford University Press, 2020

Quantum Space: Loop Quantum Gravity and the Search for the Structure of Space, Time, and the Universe, Oxford University Press, 2018

Mass: The Quest to Understand Matter from Greek Atoms to Quantum Fields, Oxford University Press, 2017

Origins: The Scientific Story of Creation, Oxford University Press, 2015

Farewell to Reality: How Fairy-tale Physics Betrays the Search for Scientific Truth, Constable, London, 2013

Higgs: The Invention and Discovery of the 'God Particle', Oxford University Press, 2012

The Quantum Story: A History in 40 Moments, Oxford University Press, 2011, re-issued 2015

Atomic: The First War of Physics and the Secret History of the Atom Bomb 1939–49, Icon Books, London, 2009, re-issued 2015 (shortlisted for the Duke of Westminster Medal for Military Literature, 2010)

A Beginner's Guide to Reality, Penguin, London, 2005

Beyond Measure: Modern Physics, Philosophy and the Meaning of Quantum Theory, Oxford University Press, 2004

Perfect Symmetry: The Accidental Discovery of Buckminsterfullerene, Oxford University Press, 1994

The Meaning of Quantum Theory: A Guide for Students of Chemistry and Physics, Oxford University Press, 1992

PROLOGUE

An Irresistible Longing to Understand the Secrets of Nature

It's probably not unreasonable to say that theoretical physics attracts particular kinds of people to work on it. This is a discipline that demands an agile, creative mind and a certain facility with abstruse concepts and dense, complex mathematics, so a degree of self-selection can be expected. A general lack of desire for material wealth is also useful. But if we're dealing with a physics perched right on the edge of our understanding of the nature of reality and physical existence, then we must admit that there's a further characteristically human trait that can often be helpful.

Theoretical physics *loves* a rebel.

Put it this way. You don't get the opportunity to transform our understanding of the very fabric of space and time; you don't get to turn the world upside-down and subvert our cosy notions of the larger universe if you're inclined to worry about what other people will think.

Many rebels come to theoretical physics seeking a refuge, a safe haven from the perceived injustices and unpredictability of human affairs and the social disappointments of youth. They come seeking a place where their instincts are more likely to be appreciated as, unlike many other walks of life, rebellion in science is not only encouraged, it is *necessary*.

At Walnut Hills High School in Cincinnati, Ohio, the sixteen-year-old Lee Smolin was principally interested in revolutionary politics, rock stardom, mathematics, architecture, and his girlfriend, not necessarily in this order or with this priority. His teachers had advised him that he wasn't smart enough to take the advanced track in mathematics and, to prove them wrong, in a singular act of rebellion he completed the three-year advanced course in just a year. Not everyone's idea of radicalism in action, perhaps, and not as subversive as rock music or publishing an underground newspaper, but Smolin discovered that 'it was almost as much fun'.[1]

His interest in architecture was kindled when, in the eleventh grade, he invited the eccentric architect and system theorist Richard Buckminster Fuller to speak at the school. A fascination with Fuller's geodesic domes led him to a branch of mathematics called tensor calculus. Books on tensor calculus led him to Einstein's theories of relativity, and to Einstein himself.

Smolin's world crumbled at the beginning of his senior year. His rock band had split, his girlfriend had left him, and his political revolution had failed to come to pass. He had flunked chemistry, and a perceived lack of aptitude meant that he had been refused admission to the physics class. He decided to drop out of high school altogether.

It was therefore in the public library that he would find the book that would change his life. It was called *Albert Einstein: Philosopher-Scientist*, edited by Northwestern University philosopher Paul Arthur Schilpp, and first published in 1949. The book opens with a chapter of 'Autobiographical Notes', written by the 67-year-old Einstein as 'something like my own obituary'.[2] His words spoke directly to the disillusioned Smolin.

Einstein wrote of the 'nothingness of the hopes and strivings which chases most men restlessly through life'. As a young man he had himself 'soon discovered the cruelty of that chase, which in those years was much more carefully covered up by hypocrisy and glittering words than is the case today'. Rejecting any solace that might be found in organized religion, Einstein had instead found comfort in physics:[3]

> Out yonder there was this huge world, which exists independently of us human beings and which stands before us like a great, eternal riddle, at least partially accessible to our inspection and thinking. The contemplation of this world beckoned like a liberation, and I soon noticed that many a man whom I had learned to esteem and admire had found inner freedom and security in devoted occupation with it.

Smolin decided to become a theoretical physicist later that evening. Like Einstein, he was 'motivated by an irresistible longing to understand the secrets of nature'.[4] '[I]t occurred to me then and there that if I could do nothing else with my life, perhaps I could do that.'[5]

It was not an entirely auspicious decision. He had already been accepted to study architecture at Hampshire College, a radical liberal arts college in Amherst, Massachusetts, and he now scrambled to switch subjects. But he was not totally unprepared. His mother, a Professor of English at the University of Cincinnati, helped enroll him on a graduate course on general relativity, taught at the university by Paul Esposito. This was his first physics course.

He also spent the hot summer months between school and college in Los Angeles working as a sheet metal apprentice at Van Nuys Heating and Air Conditioning, reading about basic physics, relativity, and quantum mechanics in his spare time.

Carlo Rovelli's journey to theoretical physics took place on a different continent, in a different language, and differs in its details. Yet it shares some remarkable similarities.

He, too, had come to have little faith in a world organized by adults in ways that seemed far from just and right. As he grew up in Verona, in northern Italy not far from Venice, he railed against the creeping nostalgia for fascism that had leached into all parts of provincial society. He clashed frequently with his teachers and rebelled against the authority of the classical lyceum, his upper secondary school, teaching basic subjects in preparation for university. He also needed to escape from his own family. A mother's love for her only child is comforting, but it can also be stifling.[6] Rovelli needed to breathe.

He read voraciously on politics, sociology, and science, and devoured novels and poetry. At the age of twenty he set off on a nomadic quest around the world in search of truth. On his travels he acquired a strong sense of liberty; he learned how to take his life in his own hands and follow his dreams. But by putting some distance between himself and the place that represented everything he had come to resent, he began to see things a little differently. There was still plenty to be angry about, but he began to realize that there were also rich possibilities for learning back in Italy. And he was also missing his Italian girlfriend.

On his return he enrolled to study for a degree in physics at the University of Bologna, the world's oldest, founded in 1088. This was more accident than design. At school he had demonstrated some capability in physics and mathematics but his first love was philosophy. He had chosen not to enroll for a philosophy degree because he simply didn't trust established educational institutions to treat philosophical problems with the importance and seriousness the young idealist demanded.

Bologna is a city famed for its art, culture, and historic architecture—notably its red-tiled roofscape, reflecting the colour of its communist politics. It suited him well. During his time as a student he made common cause with a like-minded community, one which embraced a post-hippy counterculture. The group experimented with mind-altering drugs, and different ways of living, and loving, as they tended their goat, Lucrezia. They dreamt of a peaceful, cultural revolution that would make the world a better place.

Despite the distractions of commune-style living, Rovelli had no problem maintaining his focus on physics. He would become so absorbed in study that he would remain blissfully unaware of everything else going on around him. One day, a builder arrived to demolish an interior wall in the dilapidated house in which they were living. This took several hours of noisy effort. Rovelli was working in the room, sitting just a few metres from the wall in question. When asked if the builder had disturbed him, he looked up from his books and asked: 'What builder?'[7]

In February 1976 he joined the group that established Radio Alice, a free radio station which provided: an 'open microphone for everybody, where experiences and dreams were exchanged'.[8] Topics included labour protests and political analysis, poetry, yoga, cooking, declarations of love, and music by Beethoven and Jefferson Airplane.

This was one of the defining periods of Rovelli's life, but as the dream faded he learned that 'one does not change the world so easily'.

Confused and greatly disillusioned, Rovelli now had to come to terms with the challenge of deciding what to do with the rest of his life. The timing was perhaps fortuitous. He had chosen to learn physics because he had to study something (other than philosophy), and he preferred to postpone the call to obligatory military service. But in the third year of his degree, he was at last exposed to the conceptual revolutions that had shaken physics earlier in the twentieth century. In quantum mechanics and in Einstein's relativity he would find the places where physics and philosophy not only collide, they become barely distinguishable.

Once again, Einstein provided inspiration. Shortly after completing work on relativity, Einstein wrote a popular account of his theories. He called it a 'booklet'. It was first published in German in the spring of 1917, entitled *Relativity: The Special and the General Theory (A Popular Exposition)*. He wasn't entirely satisfied with the result, and later joked that although the cover described the book as 'generally understandable', it was in fact *'gemeinunverständlich'* (generally not understandable).

The book was nevertheless enormously successful, and went through many editions, translations, and reprintings. Along the way it picked up a series of appendices as readers (and publishers) demanded a little more clarity of explanation of the mathematics, and as the observational and experimental evidence in support of relativity accumulated.

In 1953 (when Einstein was 74) he penned a fifth appendix entitled 'Relativity and the Problem of Space'. This is quite different in style from the others and contains some deep philosophical observations on the nature of space and time. It represents the result of almost fifty years of further reflection made towards the end of his life. Einstein died two years later.

In this appendix Einstein addressed questions that had teased the intellects of philosophers for centuries. 'It is indeed an exacting requirement', he wrote, 'to have to ascribe physical reality to space in general, and especially to empty space. Time and again since remotest times philosophers have resisted such a presumption.'[9]

That was it. Rovelli was captivated. This kind of physics spoke to him of 'the possibility of not giving up the desire for change and adventure, to maintain my freedom of thinking and to be what I am'.[10]

Neither of them knew it yet, but their passion for adventure in the search for the secrets of nature would eventually bring Smolin and Rovelli together, in one of the most productive and pleasurable of modern-day scientific collaborations.

To appreciate what these theorists have achieved in a collaboration spanning thirty years, we must first understand what they learned as students about two of the greatest theories of physics ever devised—relativity and quantum mechanics—and the dark secret that has kept these theories apart.

PART I
FOUNDATIONS

1

THE LAWS OF PHYSICS ARE THE SAME FOR EVERYONE

It's not hard to understand why Smolin and Rovelli would be drawn to the revolutions in scientific thinking inspired by Einstein. As they listened to their teachers in class, read diligently, and worked their way through the classic textbook problems, their minds were opened upon a landscape of extraordinary possibility.

They found themselves asking fundamental questions about the nature of the seemingly obvious—space and time—the very fabric of our physical reality. Despite familiar appearance, Einstein had shown that the answers to these questions are *not* obvious. He had shown that it is possible to subvert authority and overcome prejudice in pursuit of a deeper and more profound truth. He had set out on his path to revolution at the age of just 26 and, although his legacy is virtually without parallel in the history of science, it was clear to the young students that his work was unfinished. There was one final step that had yet to be taken.

Einstein opens Appendix 5 of *Relativity* with the following observation: 'It is a characteristic of Newtonian physics that it has to ascribe independent and real existence to space and time as well as to matter'.[1] The so-called 'classical' system of physics that the English mechanical philosopher Isaac Newton had helped to construct in the seventeenth century, some two hundred years before Einstein, demands a fabric of *absolute* space and time. This notion appears so consistent with ordinary experience that anyone unfamiliar with relativity will likely never give it a second thought.

But there are real philosophical (and, as it turns out, very practical) reasons why we should reject this notion completely.

An absolute space forms a kind of 'container' within which some sort of mysterious cosmic metronome marks absolute time. This is a container within which actions impress forces on matter and things happen. But if we were somehow able to take all the matter out of the universe, we are obliged to presume that the empty container would remain, and the metronome would continue to lay down its cosmic click-track. There would still be 'something'.

But what, exactly? There's a logic that suggests that everything there is really ought to exist *within* the universe, kind of by definition. But the notions of absolute space and time imply the opposite—that the universe instead exists within the container. If we push this logic a little further, we can imagine a vantage point from which we could look down on the entire universe: a 'God's-eye view' of all creation.

We could just shrug our shoulders at this point and argue that, grand philosophical (and theological) implications aside, absolute space and time at least appear to be consistent with our everyday experience. We're generally able to find things in the places we left them. We always follow the same route to and from work. Our days always start in the mornings. Surely, these are unassailable absolutes of our physical reality?

But even this isn't true. A moment's reflection will tell us that, despite superficial appearances, we only ever see objects moving towards or away from each other, changing their relative positions. This is *relative* motion, occurring in a space and time that are in principle defined only by the relationships between the objects themselves. Newton was obliged to acknowledge this in what he called our 'vulgar' experience.

So, we might think to reduce our vulgarity and impose some order by using a coordinate system (based, for example, on three spatial dimensions, which we define with the aid of coordinate axes, labelled x, y, and z) and by noting that an object in this place at this time moves to this other place a short time later. That's better. Or, at least, this is starting to sound a little more scientific.

But don't get too comfortable. Because we now must admit that any such coordinate system is entirely arbitrary.

We measure places on Earth relative to a different kind of coordinate system, of latitude and longitude, defined by the shape and size of our planet. We measure time relative to a system based on the orbital motion of the Earth around the Sun and the spin motion of the Earth as it turns on its axis. These systems might seem to be perfectly 'natural' choices, but they are natural only for us Earthbound human beings and we cannot escape the simple truth that they are also quite arbitrary. Systems of coordinates like x, y, and z, latitude and longitude, and so on define so-called *frames of reference* within which we can locate objects and see things happening.

We can go further. An object in uniform motion in a straight line appears to move from here to there. But what, exactly, is moving? Is it the object, travelling from here to there at a certain speed? Or is the object actually stationary, and 'there' is moving 'here' at this same speed?

Fans of J. R. R. Tolkein's *The Lord of the Rings* may remember Pippin's experiences, sat before Gandalf on the back of Shadowfax, riding in haste to Minas Tirith: 'As he fell slowly into sleep, Pippin had a strange feeling: he and Gandalf were still as stone, seated upon the statue of a running horse, while the world rolled away beneath his feet with a great noise of wind'.[2]

In such examples of uniform motion, there is in principle no observation or measurement we can make that will tell us who or what is moving. Of course, simple logic dictates that it is Shadowfax that is galloping on a stationary Middle-earth, but there's no escaping the rather stubborn fact that we can't actually *prove* this.

Such uniform motion is entirely relative, and physicists define it in the context of so-called *inertial* frames of reference. We conclude from all this that there can be no absolute coordinate system of the universe, no absolute or ultimate inertial frame of reference, and therefore no absolute motion. There is no 'God's-eye view'.

Any concept that is not accessible to observation or experiment in principle, a concept for which we can gather no empirical evidence, is typically regarded to be *metaphysical* (meaning literally 'beyond physics'). Why, then, did Newton insist on a system of absolute space and time, a metaphysical system we can never directly experience? Because by making this assumption he found that he could formulate some relatively simple—and very highly successful—*laws of motion*.

Success breeds a certain degree of comfort, and a willingness to overlook the sometimes grand assumptions or pre-commitments on which theoretical descriptions are based. Nevertheless, towards the end of the nineteenth century a growing and vociferous empiricist philosophy—one which rejected completely all metaphysical constructions and sought to exclude them from science—was shifting the weight of scientific opinion.

Momentum was building, but then Scottish physicist James Clerk Maxwell threw a hefty spanner in the works.

Confronted by compelling experimental evidence for deep connections between the phenomena of electricity and magnetism, over a ten-year period from 1855 Maxwell published a series of papers which describes these in terms of two distinct, but intimately linked, electric and magnetic *fields*. We denote such fields by drawing 'lines of force' stretching (by convention) from positive to negative, or from north pole to south pole (see Figure 1). Such fields are not creatures of a fertile imagination: we can *feel* the magnetic field when we try to push the north poles of two bar magnets together.

But this is now no longer all about material objects moving about in a three-dimensional space in a one-dimensional time. Maxwell's electromagnetic field equations tell of a very different kind of physics. A magnetic field is felt in the 'empty' space around the magnet (we can quickly verify that the field persists in a vacuum—unlike sound, it doesn't require air to 'carry' it). In fact, Maxwell's equations can be manipulated in such a way that they quite clearly describe the motions of *waves*.

This gelled quite nicely with a growing body of experimental evidence in support of a wave theory of light, and light is just one form of electromagnetic radiation. Maxwell's equations can be further manipulated to calculate the speed of electromagnetic waves travelling in a vacuum. It turns out that the result is precisely the speed of light, to which we give the special symbol c.

But it's rare in science (as indeed it is in life) that such a moment of clarity doesn't have to be paid for with confusion elsewhere. The wave nature of electromagnetic radiation now seemed to be clear and unarguable. But then physicists had to admit that these must be waves *in* something.

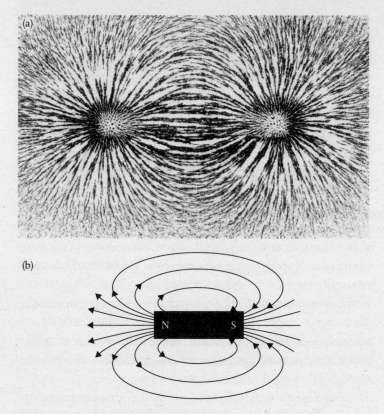

Figure 1. (a) Iron filings sprinkled on a sheet of paper held above a bar magnet reveal the 'lines of force' of the magnetic field stretching between the north and south poles. This pattern is shown schematically in (b). By convention the lines of force 'flow' from north to south.

We throw a stone into a lake and watch as the disturbance ripples across the surface of the water. The waves caused by this action are clearly waves in a 'medium'—the water in this case. There could be no escaping the conclusion. Electromagnetic waves had to be waves in some kind of medium. Maxwell himself didn't doubt that electromagnetic waves must move through the *ether*, a purely hypothetical, tenuous form of matter thought to fill all of space.

And here's the confusion, the price to be paid. All the evidence from experimental and observational physics suggested that if the ether really exists, then it couldn't be participating in the motions of observable objects. The ether *must* be stationary. If the ether is stationary, then it is also by definition absolute: it fills precisely the kind of container demanded by an absolute space. A stationary ether would define the ultimate inertial frame of reference.

Hmm.

But the problem is now subtly different. Newton required an absolute space that sits passively in the background and which, by definition, we can never experience. Now we have an absolute space that is supposed to be *filled with ether*. That's a very different prospect.

So here's a thought. If the Earth spins on its axis in a stationary ether, then we might expect there to be an *ether wind* at the surface (actually, an ether drag, but the consequences are the same). The ether is supposed to be very tenuous, so we wouldn't expect to feel this wind like we feel the wind in the air. But, just as a sound wave carried in a high wind reaches us faster than a sound wave travelling in still air, we might expect that light travelling in the direction of the ether wind should reach us faster than light travelling against this direction. A stationary ether suggests that the speed of light should be different when we look in different directions.

Any differences were expected to be very small, but nevertheless still measurable with late-nineteenth-century optical technology. But in 1887 American physicists Albert Michelson and Edward Morley could find no differences. Within the accuracy of their measurements, the speed of light was found to be constant, independent of direction. Their result suggests that there is actually no such thing as a stationary ether.

It's just this kind of conundrum that brings science to life. Newton's laws of motion demand an absolute space and time that we can't experience or gain any empirical evidence for. Maxwell's electromagnetic waves demand a stationary ether to move in, but experiment tells us that there can be no such thing. What to do?

At this point in stepped a young 'technical expert, third class' working at the Swiss Patent Office in Bern. Fed on a diet of physics and empiricist philosophy, in 1905 Einstein judged that the solution required a firmly practical and pragmatic approach in which the 'observer' takes centre stage. Here 'observer' doesn't necessarily mean a human observer. It means that, to understand the physics correctly, we must accept that this is physics as seen from the perspective of someone or something that is observing or making measurements, with a ruler and a clock.

Of course, such an observer is implicit in the physics of Newton. But Newton's laws are formulated as though the observer is somehow 'outside' of the reality in which all the action is taking place (hence, 'God's-eye view'). Einstein put the observer back into the thick of it, *inside* the reality that is being observed.

Einstein began by stating two basic principles. The first, which he called the *principle of relativity*, says that observers who find themselves in relative motion at different (but constant) speeds *must* make measurements that conform to the laws of physics. Put another way, the laws of physics must be the same for everyone, irrespective of how fast they're moving relative to what they're observing (or the other way around). This is, surely, what it means for a relationship between physical properties to be a 'law'.

Smolin and Rovelli, whose aspirations for political revolution had broken so disappointingly on the rocks of human intransigence, would have appreciated this statement when they encountered it for the first time. At least in the world of physics, true democracy reigns.

The second principle relates to the speed of light. In Newton's mechanics, speeds are simply additive. An object rolling along the deck of a ship as the ship ploughs across the Atlantic Ocean is moving with a total speed given by the speed of the roll along the deck plus the speed of the ship. But light doesn't obey this rule. The conclusion drawn from the Michelson–Morley experiment is that light always travels at the same speed. The light emitted from a flashlight moves away at the speed of light, c. Light from the same flashlight lying on the deck of the ship still moves at the speed of light, not c plus the speed of the ship.

Instead of trying to figure out *why* the speed of light is constant, Einstein simply accepted this as an established fact and proceeded to work out the consequences.

The speed of light is so incredibly fast compared with the 'everyday' speeds of objects with which we're more familiar. Normally, this means that what we see appears simultaneously with what happens. This happens over here, and we see this 'instantaneously'. That happens over there shortly afterwards, and we have no difficulty in being able to order these events in time: this first, then that. Einstein was asking a very simple and straightforward question. However it might appear to us, the speed of light is *not* infinite. If it actually takes some time for light to reach us from over here and over there, how does this affect our observations of things happening in space and in time?

Einstein discovered that one immediate consequence of a fixed speed of light is that there can be no such thing as absolute time.

Suppose you observe a remarkable occurrence. During a heavy thunderstorm you see two bolts of lightning strike the ground simultaneously, one to your left and one to your right (see Figure 2). You're standing perfectly still, so the fact that it takes a small amount of time for the light to reach you is of no

Figure 2. The stationary observer in (a) sees the lightning bolts strike simultaneously, but the observer in (b), who is moving at a considerable fraction of the speed of light, sees the right-hand bolt strike first.

real consequence. Light travels very fast so, as far as you're concerned, you see both bolts at the instant they strike.

However, I see something rather different. I'm moving at very high speed—half the speed of light, in fact—from left to right. I pass you just as you're making your observations. Because I'm moving so fast, by the time the light from the left-hand bolt has caught up with me I've actually moved quite a bit further to the right, and so the light has a little further to travel. But the light from the right-hand bolt has less ground to cover because I've now moved closer to it. The upshot is that the light from the right-hand bolt reaches me first.

You see the lightning bolts strike simultaneously. I don't. Who is right?

We're both right. The principle of relativity demands that the laws of physics must be the same for everyone, irrespective of the relative motion of the observer and, like Pippin riding on Shadowfax, we can't use physical measurements to tell whether it is you or me who is in motion.

We have no choice but to conclude that there is no such thing as absolute simultaneity. There is no definitive or privileged inertial frame of reference in which we can declare that these things happened at precisely the same time. They may happen simultaneously in this frame or they may happen at different times in a different frame, and all such frames are equally valid. Consequently, there can be no 'real' or absolute time. We perceive events differently because time is relative.

You may already be familiar with the consequences of this relativity, which later became known as 'special' relativity because it doesn't deal with objects that are accelerating.* An observer moving relative to a series of events will measure these to unfold in a time that is longer (time is dilated) when compared with the measurements of a stationary observer. The length of an object moving relative to a stationary observer will appear to contract compared with the measurements of an observer riding on the object.

The extent of time dilation and distance contraction depend on the ratio of the relative speed of the observer and the speed of light. These only become noticeable when the relative speed is close to that of light. A stationary observer won't notice the length of your car contracting, no matter how fast you drive past.

This is all a bit disconcerting, and it's tempting to slip back into older, more comfortable ways of thinking. If this is all about observation and measurement at speeds close to that of light, then surely this is just a matter of perspective and perception? From the perspective of this inertial frame of reference, time *appears* to slow down and distances *appear* to contract. Surely, time doesn't really slow down and distances don't really contract?

* But stay tuned.

Ah, but they do. Space and time are relative, not absolute, and there is therefore no unique or 'correct' perspective which will give us absolute measures of distance and time. The consequences are very practical. To be fair, it's hard to gather experimental evidence for the effects of distance contraction, but the effects of time dilation can certainly be measured.* If we put an atomic clock on a plane and fly it from London to Washington, DC and back, we find that the clock loses 16 billionths of a second compared with a stationary clock left behind at the UK's National Physical Laboratory. This is due to the fact that time slows down aboard the plane as it crosses and re-crosses the Atlantic.[3]

This is a lot to take in, and the consequences are pretty mind-blowing. The young Rovelli realized that special relativity does not sit well with the notion of a single 'present', defined everywhere. In many ways the present is an illusion, just as the flat Earth is an illusion wrought from our inability to perceive the curvature of the Earth from our vantage point on its surface. If we were somehow able to perceive time in billionths of a second, we might realize that 'saying "here and now" makes sense, but that saying "now" to designate events "happening now" throughout the universe makes no sense.'[4] Trying to establish an absolute basis for time-ordering events unfolding in the universe is as futile as trying to discover what lies north of the North Pole.

Einstein thought long and hard about these consequences, and later in 1905 published a short addendum to his paper on relativity. He applied this same logic to an object emitting two bursts of light of equal energy in opposite directions, such that the object is not diverted from its straight-line path. He deduced that the

* Simply because distance contraction happens on scales at which we must consider that other great theory of the twentieth-century—quantum mechanics—and this confuses things quite a bit.

total energy carried away by the light bursts is measured to be *larger* when observed from an inertial frame of reference moving relative to the object, just as time is dilated.

But then there is a law of physics that says that energy *must* always be conserved. Energy cannot be created or destroyed. So, if the energy carried away is measured to be larger, where then does this extra energy come from? We might instinctively assume that the object must slow down, losing some of its energy of motion—called kinetic energy—which is somehow transferred to the bursts of light. But this is not what Einstein discovered. He found that the energy does indeed come from the object's kinetic energy, but the object doesn't slow down. The energy comes instead from the *mass* of the object, which falls by an amount given by $m = E/c^2$.

Einstein concluded:[5]

> If a body emits the energy [E] in the form of radiation, its mass decreases by [E/c²]. Here it is obviously inessential that the energy taken from the body turns into radiant energy, so we are led to a more general conclusion: The mass of a body is a measure of its energy content.

Today we would probably rush to re-arrange this expression to give the iconic formula $E = mc^2$.

At the time of its publication in 1905 the special theory of relativity was breathtaking in its simplicity—the mathematics in it isn't all that complicated—yet it is profound in its implications. As young students Smolin and Rovelli marvelled at the logic and were fascinated by the conclusions.

But if Newton had been watching over Einstein's shoulder, he might still have indulged a little smile.

As I mentioned earlier, Einstein's theory is 'special' because it deals only with systems in uniform motion. It does not—it cannot—deal with systems undergoing acceleration. Although we might be prepared to admit the relativity of uniform motion in a straight line, anyone who has ever ridden a roller coaster will tell you that acceleration is something that we *feel*. Pippin had no sense of his own uniform motion riding on the back of Shadowfax, but when we're subject to a sudden *change* of speed or direction, or when we find ourselves spinning around in a circle, we *know* it.

But acceleration relative to what? Rotation relative to what? Despite the success of the special theory, Einstein hadn't yet completely closed the door on absolute space and time.

There's more. In addition to his laws of motion, Newton had also derived a law of universal gravitation. This states that objects experience a force of gravity that is proportional to their masses and inversely proportional to the square of the distance between them—we multiply the masses and divide by the distance-squared.

This was another great success, but it also came with another hefty price tag. Newton's force of gravity is distinctly different from the kinds of forces involved in his laws of motion. The latter forces are *impressed*; they are caused by actions involving physical contact between the object at rest or moving uniformly and whatever it is we are doing to change the object's motion.

Newton's gravity works differently. The force of gravity is presumed to pass instantaneously between the objects that exert it, through some kind of curious action-at-a-distance. It was not at all clear how this was supposed to work. Critics accused him of introducing 'occult forces' in his system of mechanics.

Newton had nothing to offer. In a general discussion (called a 'general scholium'), which he added to the 1713 second edition of his famous work *Mathematical Principles of Natural Philosophy*, he wrote: 'I have not been able to discover the cause of those properties of gravity from phenomena, and I frame no hypotheses.'[6]

Because Newton's force of universal gravitation is supposed to act *instantaneously* on objects no matter how far apart they might be, this classical conception of gravity is completely at odds with special relativity, which denies that the influence of any force can be transmitted faster than the speed of light.

Special relativity could not cope with acceleration and it could not accommodate Newton's force of gravity. Einstein still had plenty of work to do.

2

THERE'S NO SUCH THING AS THE FORCE OF GRAVITY

Newton was all too aware that he was vulnerable on the question of absolute space, but acceleration (and, particularly, rotation) was his secret weapon.* In an attempt to pre-empt his critics, he devised a thought experiment to show how the very possibility of rotational motion proves the existence of absolute space. This is Newton's famous 'bucket experiment'.

In his *Autobiographical Notes*, Einstein mentions this only in passing: 'First in line to be mentioned is Mach's argument, which, however, had already been clearly recognized by Newton (bucket experiment).'[1]

Einstein doesn't refer to Newton's bucket in Appendix 5 of *Relativity*, but he does credit Austrian Ernst Mach as the only

---

* Acceleration is the rate of change of speed with time. We tend to think of this as involving a change in the magnitude of the speed (for example, when we accelerate in a car from 0 to 60 miles per hour). But speed is a *vector* quantity—it is described in terms of both magnitude *and* direction. If we keep the same speed but rapidly change direction, this is still acceleration. So rotation, in which the direction of motion is constantly changing, is a very particular kind of acceleration.

physicist 'who thought seriously of an elimination of the concept of space, in that he sought to replace it by the notion of the totality of the instantaneous distances between all material points. (He made this attempt in order to arrive at a satisfactory understanding of inertia.)'[2]

Newton's thought experiment runs something like this. We tie one end of a rope to the handle of a bucket and the other end around the branch of a tree, so that the bucket is suspended in mid-air. We fill the bucket three-quarters full with water. Now we turn the bucket so that the rope twists tighter and tighter. When the rope is twisted as tight as we can make it, we let go and watch what happens (Figure 3).

The bucket begins to spin around as the rope untwists. At first, we see that the water in the bucket remains still. Then, as the bucket picks up speed, the water itself starts to spin and its surface becomes concave—the rotational motion appears to exert a centrifugal force which pushes the water out towards the circumference and up the inside of the bucket. Eventually, the rate of spin of the water catches up with the rate of spin of the bucket, and both spin around together.

Figure 3. Newton's bucket. As the rope untwists, the bucket rotates and the water inside is driven up the inside, forming a concave shape.

In the *Mathematical Principles*, Newton wrote:[3]

> This ascent of the water [up the inside of the bucket] shows its endeavour to recede from the axis of its motion; and the true and absolute circular motion of the water, which is here directly contrary to the relative, discovers itself, and may be measured by this endeavour.

This is such an ordinary or everyday kind of observation that it would seem to prove nothing, let alone the existence of absolute space. But the logic is quite compelling. The water pushed up the inside of the bucket is obviously moving, and we accept that this motion must be *either* absolute *or* relative. The water continues to be pushed up the inside of the bucket as its rate of spin *relative to the bucket* changes, and it remains in this state when the water and the bucket are spinning around at the same speed. Newton argued that the origin of this behaviour cannot therefore be traced to the motion of the water relative to the bucket. If this motion isn't relative, then it must be absolute. And if absolute motion is possible, then absolute space must exist.

Einstein was aware of the flaw in Newton's logic, but the counter-argument takes some swallowing. Many years later it was argued that Newton had neglected to consider the bigger picture. Yes, the behaviour of the water in the bucket cannot be explained by considering only its motion relative to the bucket. But it can potentially be explained by considering its motion *relative to the rest of the universe*.

Remember, if all motion (including rotation) is relative, then there is in principle no observation or measurement we can make which tells us who or what is moving. This is what 'relative motion' means. Newton's argument fails if it turns out that we can't distinguish between the situation in which the bucket is spinning relative to the rest of the universe, and the situation in which the rest of the universe is spinning relative to the stationary bucket.

Of course, this leads us to the rather bizarre conclusion that spinning the entire universe around a stationary bucket would somehow exert a centrifugal force on the water inside it. We're left to ponder just how that might work.

As is evident from Einstein's comments, this counter-argument is most closely associated with the physicist (and arch-empiricist) Mach, and is sometimes referred to as *Mach's principle*.[4] To eliminate absolute space entirely, Einstein needed to find a situation in which an observer experiencing acceleration wouldn't be able to tell who or what was being accelerated.

Now, all our experience on Earth suggests that acceleration (or inertia—a measure of an object's *resistance* to a change in its state of motion) is something that we feel directly and is therefore undeniable. But what happens if we find ourselves falling freely in outer space?

We can't know what Einstein was thinking, but we do know that on an otherwise perfectly ordinary day at the Swiss Patent Office in November 1907, Einstein had what he later called his 'happiest thought'.[5] He had by this time received a promotion, to 'technical expert, second class'. As he later recounted: 'I was sitting in a chair in my patent office at Bern. Suddenly a thought struck me: If a man falls freely, he would not feel his weight. I was taken aback. This simple thought experiment made a deep impression on me.'[6]

In free fall we feel neither acceleration nor gravity. From this very simple intuition, Einstein realized that our experience of acceleration is precisely the same as our experience of gravity. They are one and the same thing. He called it the *equivalence principle*. This meant that solving the problem of acceleration in relativity might also solve the problem of Newton's gravity. Perhaps there were not two distinct problems to be solved, after all. Smolin recalls reading Einstein's 1907 paper on the equivalence principle whilst riding the New York City subway, and 'getting it'.[7]

Einstein had found the equivalence principle but he was unsure what to do with it. In any case his life changed quite dramatically towards the end of 1907, as his growing reputation allowed him to establish the beginnings of an academic career, at universities first in Zurich and then in Prague. He acquired many teaching and administrative responsibilities and was drawn to other problems in physics. It would take him another five years to figure out that the equivalence principle implies another extraordinary connection, between gravity and *geometry*.

But the geometry of what, exactly? The answer to this question was supplied in 1908 by Hermann Minkowski, Einstein's former mathematics teacher at the Zurich Polytechnic. In special relativity time dilates and distances contract, but Minkowski realized that it is possible to combine space and time together in such a way that these effects compensate. The result is a four-dimensional *spacetime*, sometimes called a *spacetime metric*.

Einstein later took pains to point out that this kind of 'four-dimensional' view of space and time was not particularly new. An event taking place in Newton's classical physics of everyday requires four numbers to describe it completely: three spatial coordinates x, y, and z and a time t. But in Newton's physics time is treated quite distinctly and independently of space. In Minkowski spacetime, time (t, in seconds) is multiplied by the speed of light (c, in metres per second) and so the product ct has the same units as a spatial dimension (metres), just like x, y, and z. In Minkowski spacetime, time is treated on an 'equal footing'.

If gravity is equivalent to acceleration, then Newton's (probably apocryphal) experience of the apple falling from the tree in the garden at Woolsthorpe Manor can be viewed in two distinctly different, but physically equivalent, ways. We can imagine that the force of gravity somehow reaches up and pulls the apple down to the ground. Alternatively, we can imagine that the ground accelerates upwards to meet the apple. Both are equivalent, but the latter perspective can only be applied if we regard the Earth to be flat. Of course, the Earth is curved, and we can't ignore the perfectly legitimate experiences of all the people on the other side of the world.

Einstein began to understand that the problem lay in the nature of spacetime itself. Minkowski spacetime is 'flat', or Euclidean, named for the famed Greek mathematician Euclid of Alexandria. In school we learn that the angles of a triangle add up to 180°. We learn that the circumference of a circle is 2π times its radius, and that parallel lines never meet. These are all characteristics of a flat space, and when we add a fourth dimension of time we get a flat spacetime.

As he had already done so often to great effect, Einstein once again turned the problem on its head. It is possible to demonstrate the equivalence of acceleration and gravity in a system in which a flat Earth moves through a flat spacetime. But we know that the surface of the Earth is curved. So, where does that leave spacetime?

In a flat spacetime the shortest distance between two points is obviously the straight line we can draw between them. But the shortest distance between London, England and Sydney, Australia—a distance of 10,553 miles—is not, in fact, a straight line. The shortest distance between two points on the surface of a sphere is a curved path called an *arc of a great circle* or a *geodesic*.

This was the solution that Einstein had sought. In a flat spacetime all lines are straight, so Newton's force of gravity is obliged to act instantaneously, and at a distance. But if spacetime is instead curved like an arc of a great circle, then an object moving along such a path is 'falling'. And as it falls, it accelerates.

Einstein's great insight was that spacetime is not rigid and inflexible. It is plastic. Like the 'lines of force' of a magnetic field, spacetime bends and curves this way and that, in response to the presence of mass-energy. A large object such as a star or a planet curves the spacetime around it, as shown in Figure 4, much as a child curves the stretchy fabric of a trampoline as she bounces up and down. Other objects such as planets or moons that stray too close follow the shortest path determined by this curvature. The acceleration associated with free fall along the shortest path is then entirely equivalent to an acceleration due to an apparent 'force' of gravity.

The apple falling from the tree doesn't need to be tugged down by some mysteriously instantaneous force which reaches up from the ground. It quite happily falls along the curvature of spacetime caused by the mass of the Earth and accelerates all on its own.

In May 1952, the American theorist John Wheeler at Princeton University pulled a new bound notebook from his shelf and labelled it 'Relativity I'. He was pleased to learn that he had been given the go-ahead to teach a course on relativity and thought to

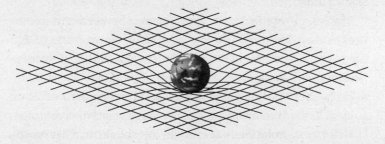

Figure 4. An object with a large mass-energy, such as the Earth, curves the spacetime around it.

get into the subject properly by writing a book about it. 'That fall, fifteen graduate students enrolled in my course,' Wheeler explained. 'It was the first time that a relativity course had been offered at Princeton—and together we worked our way through the subject, trying to get behind the mathematical formalism that had dominated the theory for decades, looking for real, tangible physics.'[8]

The book would eventually be published in 1973, entitled *Gravitation* and co-authored with Charles Misner and Kip Thorne. During his first year at Hampshire College, Smolin attended his first scientific conference, one of a series of biannual Texas Symposia on Relativistic Astrophysics, held in New York City. There he met Oxford mathematical physicist Roger Penrose and listened to talks by Stephen Hawking and American theorist Bryce DeWitt. He also met Thorne, who suggested that he obtain a copy of the new book on gravitation and study it closely. The book was not a required course text, but the following year Smolin studied it anyway.

It would prove to be an extraordinarily influential text. Running to nearly 1300 pages it is also a hefty tome, handy for those more practically minded students wishing to offer a demonstration of gravity—it will follow the curvature of spacetime and fall to the floor, landing with a very satisfying 'whump'.

Always ready with an apt name or turn of phrase, Wheeler summarized Einstein's theory of gravity some years later: 'Spacetime tells matter how to move; matter tells spacetime how to curve.'[9]

Through this insight Einstein saw that he might now be able to account for both acceleration and gravity in what would become known as the *general theory of relativity*. What this theory suggests is that there is actually no such thing as the force of gravity. Mass-energy generates a gravitational field, but this is not like a magnetic field: it is not something that exists at each and every point *in* spacetime. Rather, the gravitational field *is* spacetime.

Einstein had the idea, but he now needed to find a way to express it mathematically. This was always going to be difficult. Doing physics in four dimensions is already quite demanding, but Einstein needed a set of equations which could accommodate *any* kind of spacetime geometry. Indeed, the theory needed to accommodate spacetimes in which the angles between coordinate axes *vary* from one point to the next.

He was guided in his search by two further principles. The first is the principle of *general covariance*, essentially an extension of the principle of relativity which seeks to ensure that the laws of physics are completely independent of the choice of reference frame—inertial and accelerating—and so independent of the choice of coordinate system.*

The second is the principle of *consistency*. Einstein acknowledged that there's no faulting the accuracy of Newton's law of universal gravitation within the limits of its applicability. He therefore demanded that the law should be derivable from his field equations as a limiting case, in which spacetime can be assumed to be flat and speeds are well below light-speed.

---

* The notion of 'covariance' will prove to be very important in subsequent chapters, but this is likely to be an unfamiliar term. Remember, this is all about making sure that the underlying physics is unaffected by an arbitrary change in coordinates. If, for example, we change from coordinates measured in metres to kilometres (*multiplying* the scale by 1000), the magnitude of a physical quantity such as speed must be *divided* by 1000—a speed of 1000 metres per second becomes 1 kilometre per second. Speed is therefore considered to be a *contravariant* vector—we multiply the scale and divide the magnitude (and vice versa). In contrast, covariant vectors are like gradients. For example, a temperature gradient of 0.001°C per metre becomes 1°C per kilometre (the temperature increases 1°C for every kilometre). For a covariant vector, when we multiply the scale by 1000 we also multiply the magnitude of the vector by 1000.

It's a common misconception that Einstein the genius physicist must also have been a great mathematician. He was not. Responding to his cry for help, his friend Marcel Grossman directed him to the work of the great German mathematician Carl Friedrich Gauss and his student, Bernhard Riemann. Einstein was introduced to the Riemann curvature tensor, the definitive mathematical expression for the curvature of a generalized geometry, in any set of coordinates. This was the connection that would later bring the sixteen-year-old Smolin from architecture to general relativity.

But the Riemann tensor didn't quite fit the bill, and as Einstein struggled to find a way to generalize it, he got 'lost' in the algebra, for two years making mistakes and following alleys that each in turn proved to be blind.

He eventually found the generalization required to make the field equations conform satisfactorily to the demands of the physics. The result is a set of equations which connect the curvature of spacetime—on the left-hand side—with the distribution and flow of mass-energy on the right-hand side. He presented his equations to the Prussian Academy of Sciences in Berlin in a series of four lectures which concluded on 25 November 1915.

From the beginning of his journey to the gravitational field equations of general relativity, Einstein was aware of four potential empirical tests.

Actually, the first is not so much a test, more the resolution of a mystery. We've known since the work of Johannes Kepler in the seventeenth century that a planet moves in an elliptical orbit around the Sun. But the planetary orbits are not exact ellipses. If they were, each planet's point of closest approach to the Sun (which is called the *perihelion*) would be fixed, the planet always passing through the same point with each and every orbit. But astronomical observations show that with each orbit the perihelion shifts slightly, or *precesses*.

Much of the observed precession is caused by the cumulative gravitational pull of all the other planets in the Solar System, effects which can be predicted using Newton's law. For the planet Mercury, lying closest to the Sun, this 'Newtonian precession' is predicted to be 532 arc-seconds per century.* However, the observed precession is rather more, about 574 arc-seconds per century, a difference of 42 arc-seconds. Though small, this difference accumulates and is equivalent to one 'extra' orbit every three million years or so.

Newton's gravity can't account for this small difference and other explanations—such as the existence of another planet, closer to the Sun than Mercury (named Vulcan)—were suggested. Astronomers searched for it in vain. Einstein was delighted to discover that the field equations of general relativity predict a further 'relativistic' contribution of 43 arc-seconds per century, due to the curvature of spacetime around the Sun in the vicinity of Mercury.† This discovery gave Einstein the strongest emotional experience of his life in science: 'I was beside myself with joy and excitement for days.'[10]

Perhaps the most famous prediction of general relativity concerns the bending of starlight passing close to the Sun. Once again, it's not the 'bending' per se that represents a novel prediction, but rather its extent. Newton's gravity suggests that light grazing the surface of the Sun should bend through 0.85 arc-seconds.‡ But the curvature of spacetime predicted by general

---

* A full circle is 360°, and an arc-minute is one-sixtieth of 1°. An arc-second is then one-sixtieth of an arc-minute. So 532 arc-seconds represents about 0.15 of a degree.

† The perihelia of other planets are also susceptible to precession caused by the curvature of spacetime, but as these planets are further away from the Sun the contributions are much less pronounced.

‡ Newton favoured a 'corpuscular' theory in which light is imagined to consist of individual particles. It's therefore not difficult to imagine why such 'bullets' of light might be affected by the gravitational pull of a large object such as a star.

relativity effectively doubles this, giving a total shift of 1.7 arc-seconds. Unlike the advance in the perihelion of Mercury, nobody had yet measured the extent to which starlight bends as it passes close by the Sun, so here was a direct test.

Einstein's prediction was famously borne out by a team led by British astrophysicist Arthur Eddington in May 1919, which reported measurements on light from a number of stars that grazed the Sun on its way to Earth during a total solar eclipse. Although few could really understand the implications (and fewer still—even within the community of professional physicists—could follow the abstract tensor mathematics), the notion of curved spacetime captured the public's imagination. Einstein was an overnight sensation.

General relativity also predicts effects arising from curved spacetime that are similar in some ways to the effects of special relativity, and Einstein had worked out the details in 1911. He deduced that time will dilate and distances will contract when measured close to a large object where the curvature of space-time is strongest. A standard clock on Earth will run more slowly than a clock placed in orbit around it.

As we learned in the last chapter, an atomic clock carried aboard a plane flying from London to Washington, DC and back loses 16 billionths of a second relative to a stationary clock left behind at the UK's National Physical Laboratory. This is time dilation associated with the *speed* of the aircraft and is an effect of special relativity. But the clock *gains* 53 billionths of a second due to the fact that gravity is weaker (spacetime is less curved) 10 kilometres above sea level, an effect of general relativity. After some small geometric adjustments the net gain in this experiment was predicted to be about 40 billionths of a second. When these measurements were actually performed in 2005, the measured gain was reported to be 39 ± 2 billionths of a second.[11]

Now, you might think that a few tens of a billionth of a second are neither here nor there in the bigger scheme of daily life on planet Earth. But without the subtle corrections required by both special and general relativity, the Global Positioning System (GPS) device used by one or more of your smartphone apps or the navigation system in your car, boat, or aircraft would quickly accumulate clock errors, giving rise to positioning errors of more than 11 kilometres in a single day.[12] You would have a real hard time tracking your run or cycle ride, or navigating your way to your destination across the ocean or through the air.

Einstein's field equations were so complicated that Einstein himself judged them impossible to solve exactly, without first making some simplifying assumptions. And yet within a year the German mathematician Karl Schwarzschild had worked out a set of solutions for a large, electrically uncharged, non-rotating spherical body, which serves as a useful approximation for slowly rotating objects such as stars and planets. Smolin chose the Schwarzschild solutions as the subject for an oral examination shortly after his arrival at Hampshire College. Alas, he had over-reached himself, and gaps in his understanding were quickly exposed. He was failed.[13]

One of the more startling features of the Schwarzschild solutions is that they predict a fundamental boundary—called the *Schwarzschild radius*. Any object compressed to a radius smaller than its Schwarzschild radius becomes so dense that it curves space-time back on itself (the Schwarzschild radius of the Earth is about nine millimetres). Nothing—not even light—can escape the pull of such an object's gravitational field. The result is a *black hole*.*

'When I was a university student,' wrote Rovelli, many years later, 'black holes were regarded as a scarcely credible implication

---

* A name popularized (though not coined) by Wheeler.

of an esoteric theory. Today they are observed in their hundreds and studied in detail by astronomers.'[14] Although black holes are obviously difficult to detect directly, there is now plenty of indirect evidence to suggest that they are fairly ubiquitous in the universe, and supermassive black holes are likely to sit at the centres of every galaxy.

Einstein was pretty cool on the idea of black holes, but in June 1916 he speculated that small fluctuations in a gravitational field would, like the ripples on the surface of a lake, appear as waves.* Such *gravitational waves* are of a very different kind to light waves and can only be produced by two large masses rotating around each other in what astronomers call a binary system, stretching and compressing the spacetime between them. It was not until the 1950s and 1960s that physicists thought they might stand a chance of actually detecting gravitational waves. On 15 September 2015 their patience was finally rewarded.

This was the date on which gravitational waves generated by the merger of two black holes was detected for the first time by an experimental collaboration called LIGO, which stands for Laser Interferometry Gravitational-Wave Observatory. LIGO actually involves two observatories, one in Livingston, Louisiana and another at Hanford, near Richland in Washington, essentially on opposite sides of the continental United States. The result was announced at a press conference on 11 February 2016. Since then, there have been announcements of the detection of further gravitational wave events by LIGO and the Virgo interferometer, located near Pisa in Italy. One of these events was triggered by the merger of two neutron stars. The 2017 Nobel Prize for physics was awarded to American physicists Barry Barish,

---

* Einstein made a catastrophic error in this paper which he put right two years later.

Kip Thorne, and Rainer Weiss for their contributions to LIGO and the observation of gravitational waves.

The successful detection of gravitational waves is not only an extraordinary vindication of general relativity; it also opens a new window on events in distant parts of the universe, one that doesn't rely on light or other forms of electromagnetic radiation to tell us what's happening.

When Einstein delivered the last of his lectures on general relativity at the Prussian Academy of Sciences, he believed he had finally settled the matter of absolute space and time. He wrote that the theory's principle of general covariance: 'takes away from space and time the last remnant of physical objectivity'.[15] He thus declared the defeat of the absolute and the triumph of the relative.

But now we must return to Mach's principle. If Newton's bucket is stationary and the rest of the universe is spinning around it, what then causes the centrifugal force that drives the water up the inside?

The answer is truly breathtaking. We expect that the stationary water would be set in motion because all the mass-energy in the universe collectively drags spacetime around with it as it spins. This is an effect first deduced from general relativity in 1918 by Austrian physicists Josef Lense and Hans Thirring, known variously as *frame-dragging* or the Lense–Thirring effect. The possibility of frame-dragging means that there really is no measurement we can make that would tell us whether it is the water that is rotating in a stationary universe or the universe that is rotating around a stationary bucket of water. The rotational motion of the water is relative.

On 24 April 2004, an exquisitely delicate instrument called Gravity Probe B was launched into polar orbit. The satellite housed four gyroscopes whose orientations were monitored continuously as the satellite orbited the Earth. Two effects were measured. The Earth curves the spacetime in its vicinity, and this was expected to cause the gyroscopes to precess by a predicted 6606 milliarc-seconds per year in the plane of the satellite's orbit (that is, in a north–south direction).* This precession is called *geodetic drift*, a phenomenon first identified by Dutch physicist Willem de Sitter in 1916.

The second effect is frame-dragging. As the Earth rotates on its axis, it drags spacetime around with it in the plane perpendicular to the plane of the satellite orbit (in a west–east direction). This gives rise to a second precession of the gyroscopes, predicted to be 39.2 milliarc-seconds per year.

Data collection began in August 2004 and concluded about a year later. The project suffered a major disappointment when it was discovered that the gyroscopes were experiencing a substantial and unexpected wobble, due to an unforeseen build-up of electrostatic charge. These effects could be corrected using an elaborate mathematical model, but only at the cost of increased uncertainty in the final experimental result.

Analysis of the data took a further five years. The results were announced at a press conference on 4 May 2011. The geodetic drift was reported to be 6602±18 milliarc-seconds per year. The west–east drift caused by frame-dragging was reported to be 37.2±7.2 milliarc-seconds per year. The high (19 per cent) uncertainty in this last result was caused by the need to model the unexpected wobble.

Despite the uncertainty, this is another very powerful experimental vindication of general relativity.

For students learning about relativity in the late 1970s, some of these empirical proofs would await them in the future. But even

---

* A milliarc-second is one thousandth of an arc-second.

the most rebellious of students would not doubt the theory's essential correctness. Yes, the theory is complex, written in a mathematical language only a privileged few in any generation would come to understand, with a conceptual beauty that is arguably unrivalled in the history of physics. The theory casts a spell, and Smolin and Rovelli were enchanted. '[I]t is a glance towards reality. Or better, a glimpse of reality, a little less veiled than our blurred and banal everyday view of it. A reality which seems to be made of the same stuff our dreams are made of, but which is nevertheless more real than our clouded daily dreaming.'[16]

Einstein had successfully argued that spacetime is relative. It owes its existence to matter and energy. Take all the matter and energy out of the universe and there would be no empty container. There would be nothing at all.

Everything there is, is in the universe.

3

WHY NOBODY UNDERSTANDS QUANTUM MECHANICS

Twentieth-century physics was marked by not one, but two revolutions. Einstein's special and general theories of relativity taught us some new things about the nature of space and time, and connected mass and energy together to give us mass-energy. But, despite the startling implications of Einstein's most famous equation, $E = mc^2$, and the rather incredible role of mass-energy in curving spacetime, these relationships remain firmly 'classical'. As far as relativity is concerned, the stuff that exists *within* spacetime—matter (consisting of 'particles' of some kind) and radiation (consisting of 'waves')—is assumed to behave pretty much as Newton and Maxwell and their contemporaries might have imagined.

I say this because of course the second revolution in twentieth-century physics utterly transformed our understanding of the nature of matter and radiation, their relationship with energy, and the very meaning of the *m* in $E = mc^2$. This is the revolution that gave us *quantum mechanics*.

Now quantum mechanics is often presented to students in a rather dry, matter-of-fact way, as an accepted physical theory backed up by lots of hard empirical evidence. Students are told that this is how nature is, and they just need to get used to it. They're given the mathematical prescriptions which allow them to apply the theory (or, at least, answer their exam questions correctly), without necessarily being helped to understand where these prescriptions have come from or why they have the form that they do.

The early historical development of quantum theory is typically compressed into the opening minutes of a first introductory lecture. Students learn about Max Planck's 'act of desperation' which led him to suggest that electromagnetic radiation behaves *as though* it is composed of discrete bundles of energy, which he called quanta. They learn of Einstein's light-quantum hypothesis: his bold and, at the time, rather foolhardy proposal that light actually *is* composed of discrete quanta (which we now call photons). And thus it was really Einstein, not Planck, who launched the quantum revolution.

The students learn of Niels Bohr's application of quantum ideas to describe the behaviour of the electron orbiting the proton in a hydrogen atom. In this old 'planetary' model of the atom, the absorption or emission of light causes the electron to 'jump' from one orbit to another, marked by telltale lines in the resulting spectrum which march to a mathematical pattern, getting closer and closer together as the energies of the jumps increase.

They learn of Louis de Broglie's outrageous suggestion that if electromagnetic waves can behave like particles (light quanta), then perhaps particles like electrons can behave like waves, in a kind of wave–particle 'duality'. They learn about the 'atomic orbitals' derived from Erwin Schrödinger's rather beautiful wave mechanical theory of the hydrogen atom, and of Max Born's interpretation of these electron waves as 'waves of quantum probability'. They learn of Werner Heisenberg's uncertainty principle, and of Paul Dirac's successful synthesis of quantum theory and special relativity, yielding a theory which explained the rather puzzling phenomenon of electron spin and predicted the existence of antimatter, with all the same properties of matter but with opposite electrical charge.

But then there are the things they don't learn about. Many students aren't told about the frustration, anguish, and bitter shedding of tears (yes, tears) suffered by these physicists as they argued endlessly back and forth about what quantum theory *means* for our understanding of physical reality.

They don't learn about Schrödinger's insistence that 'the whole idea of quantum jumps is bound to end in nonsense'.[1] They don't learn about the highly charged debate between Bohr and Heisenberg on the interpretation of the uncertainty principle. Some might have heard of Einstein's famous remark 'God does not play dice',[2] but likely not in the context of his debate with Bohr on the interpretation of quantum probability and the consistency and completeness of quantum theory, simply one of the most marvellous debates in the entire history of science.

And they don't learn of the series of letters exchanged between Einstein and Schrödinger which led to the formulation of the famous paradox of Schrödinger's cat. Or that this paradox was intended as a tongue-in-cheek commentary on the absurdity of the prevailing orthodox interpretation of quantum theory that had been devised by Bohr, Heisenberg, and Wolfgang Pauli.

I think this is a great pity. For students tutored to this point in classical mechanics, learning about the concepts of quantum mechanics pulls the rug of comprehension firmly from beneath their feet. A less enlightened teacher with little or no respect for the history of the subject will tell those students struggling to come to terms with it to set aside their concerns and get on with the important business of using quantum theory to *calculate* things.

A good example of this 'instrumentalist' attitude is afforded by Dirac's *Principles of Quantum Mechanics*, first published in 1930. Dirac was in many ways a very singular individual. Although he made some major breakthroughs in quantum physics, he was a mathematician first, a physicist second. This much comes through his classic text, in which he adopts 'the symbolic method, which deals directly in an abstract way with the quantities of fundamental importance', and which 'necessitates a complete break from the historical line of development'.[3]

Dirac's obsession was with methodology and formalism. He had little time for philosophy, dismissing it as a 'way of talking about discoveries that have already been made', meaning that philosophy in itself was unlikely to lead to any new discoveries.[4] As Bohr, Einstein, Heisenberg, and others argued about the interpretation of quantum theory, Dirac was unmoved by such philosophical nitpicking: 'It seemed to me that the foundation of the work of a mathematical physicist is to get the correct equations...the interpretation of those equations was only of secondary importance.'[5] He was happy to leave all the wrangling about interpretation to others.

Now, it's difficult enough to learn all the mathematical prescriptions, and if it all seems bizarre and makes little sense, perhaps it's best to remember that quantum theory is also unquestionably *true* (at least insofar as anything is 'true' in science). It is, after all, the foundation on which much of contemporary physics, a lot of chemistry, and all modern electronic technology is built.

And this is what good students do. Lee Smolin studied Dirac's text in his first year at Hampshire College. Carlo Rovelli came to it by a more circuitous route. The teacher of a course on mathematical methods for physics told each of his students to pick a topic that was not included in the programme, study it, and deliver a seminar to the rest of the group. Rovelli's chosen topic had applications to quantum mechanics, and his teacher advised him to cover this also in his seminar. 'I gently said that none of the students in the class had taken the quantum mechanics course yet.'[6] Puzzled, the teacher responded: 'So what? Study it!'

So Rovelli sat down with Dirac's classic text, together with half a dozen other books, and immersed himself fully in the subject. Two weeks later he returned to his teacher and declared: 'I have learned quantum mechanics'. The teacher looked surprised and explained that he had actually been joking. He had not expected Rovelli to learn quantum theory, by himself, in just a couple of weeks.

Good students accept the formalism at face value, get their heads down, and get on with it. But for many the deep sense of unease never really goes away. It's perhaps not so hard to appreciate why the charismatic American physicist Richard Feynman once felt moved to state: 'I think I can safely say that *nobody* understands quantum mechanics.'[7]

For somebody coming to quantum mechanics for the first time, it is perhaps difficult to see what all the fuss is about. The best way to understand why even Nobel laureates struggle to come to terms with the theory is to study what, on the surface, appears to be a relatively simple phenomenon, called two-slit or double-slit interference. As Feynman explained in his *Lectures on Physics*, 'We choose to examine a phenomenon which is impossible, *absolutely* impossible, to explain in any classical way.... In reality, it contains the *only* mystery. We cannot make the mystery go away by "explaining" how it works.'[8]

Two-slit interference was discovered by physicists in the early nineteenth century. It works like this. Form a light beam consisting of a single colour or narrow range of wavelengths. Pass this through an aperture or slit in a metal plate, cut with a width which is on the order of the wavelength, and the light beyond spreads out. Physicists call this *diffraction*. The light squeezes through the slit and is detected on a far screen or photographic plate. We see that the light appears on the screen not as a sharp line with the same dimensions as the slit, but instead as a much broader, more diffuse band.

Now shine this light on two slits side by side and it will diffract through both to produce a pattern of alternating bright and dark lines, called *interference fringes* (Figure 5). When it was first discovered, this kind of phenomenon provided strong evidence for the wave theory of light. Waves squeezed through each of the slits bend around the edges as they pass through and diffract. Waves diffracted by each slit run into each other and overlap, and where wave peak meets wave peak the result is constructive interference—the waves mutually reinforce to produce a bigger peak, or a bright fringe. Where wave trough meets wave trough the result is a deeper trough. But where wave peak meets wave trough the result is destructive interference—the waves cancel each other out, giving a dark fringe.

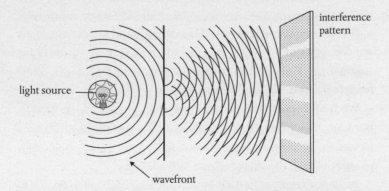

Figure 5. When passed through two narrow, closely spaced apertures or slits, monochromatic light produces a pattern of alternating light and dark fringes. These can be readily explained in terms of a wave theory of light in which overlapping waves interfere constructively (giving rise to a bright fringe) and destructively (dark fringe).

Only waves can do this kind of thing.

This all seems perfectly reasonable. But in 1905 Einstein had suggested that light might consist of 'particles' (quanta), after all. Somehow, light could behave as particles *and* waves. In 1923 French physicist Louis de Broglie took this a step further. He suggested that small particles of matter, such as electrons, might also be considered to behave like waves. Just as with light, we can form a beam of electrons, much like in an old-fashioned television tube. Now imagine that we pass this through a plate in which we've cut two small, closely spaced slits. What happens?

Our instinct might be to imagine that the electrons in the beam will pass through either one slit or the other, producing two bright lines on the screen marking where the electrons have passed through the slits. We would expect each line to be brightest in the centre, showing where most of the electrons have passed straight through the corresponding slit, becoming a little more diffuse as we move away from the centre, signalling electrons that have caught the edges of the slit and scattered on their way through.

But these experiments have been done, and this is not what we see. Instead of two bright lines characteristic of particles following straight paths through the slits, we get a two-slit interference pattern. De Broglie was right.

We might scratch our heads and shrug our shoulders at this, but let's push the experiment a little further. Let's limit the intensity of the electron beam so that, on average, only *one* electron passes through the slits at a time. What then?

What we see is, at first, quite comforting. Each electron registers as a bright dot on the screen, indicating that 'an electron struck here'. We might sigh with relief, as it seems that electrons are particles, after all. They pass—one by one—through the slits and hit the screen in a seemingly random pattern.

But wait. The pattern isn't random. As more and more electrons pass through the slits we cross a threshold. We begin to see individual dots group together, overlap, and merge, forming a 'picket fence' pattern of bright fence posts with dark gaps in-between. Eventually we get a two-slit interference pattern of alternating bright and dark fringes, as shown in Figure 6.

We're now faced with a choice. We could suppose that this wave behaviour results from some kind of statistical averaging. In this interpretation each individual electron is imagined to pass (quite logically) through one slit or the other as a self-contained elementary particle. We need further to suppose that there must be some unknown mechanism governing the subsequent behaviour of each electron that affects its path, such that—on average—it has a higher probability of hitting the screen in a location we will come to identify as a bright fringe. We can quickly discover that if we close one or the other slit or try to detect the electron passing through one or the other then we will lose the interference pattern. We just get a behaviour characteristic of particles following straight-line paths. So, whatever the mechanism is, it must depend somehow on the existence of the slit through which the electron *does not pass*, which is decidedly odd.

Alternatively, we might suppose that the wave nature of the electron is an *intrinsic* behaviour. Each individual electron behaves as a wave, passing through both slits simultaneously and interfering with itself.

Now, if this interpretation makes you feel a little uncomfortable you're really not going to like what comes next.

We imagine that an electron wave passes through both slits, interferes and moves on to hit the screen. We expect that the resulting, post-interference electron wave will exhibit an alternating pattern in the heights of its peaks and the depths of its troughs that reflect what we will eventually see as the interference pattern.

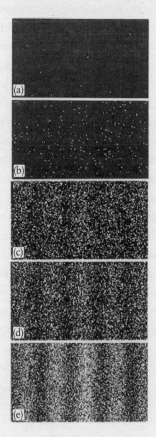

Figure 6. We can observe electrons as they pass, one at a time, through a two-slit apparatus by recording where they strike a piece of photographic film. Photographs (a) to (e) show the resulting images when, respectively, 10, 100, 3,000, 20,000, and 70,000 electrons have been detected.

But, by its very nature, the wave is *distributed* through space. It is 'delocalized'. And yet we know that the electron will be detected as a bright spot, in only *one* location on the screen. It is 'localized'.

To make sense of this we must reach for Born's interpretation of the electron wave. We presume it represents the quantum probability of 'finding' the electron. So the alternating peaks and troughs of the electron wave translate into a pattern of quantum probabilities—in this location (bright fringe) there's a higher probability of finding the electron, in this other location (dark fringe) there's a low or zero probability of finding the electron.

There's one problem with this. At the moment the electron hits the screen, it can in principle be found at *any* location where the quantum probability is greater than zero. But it is detected in only *one* location. Somehow, the electron transforms from being 'here', 'there', or 'most anywhere' to being only 'here'. This is called the *collapse of the wavefunction* (see Figure 7).

It gets worse. In quantum mechanics the collapse of the wave-function is entirely hypothetical. It is *assumed* to happen, as a way

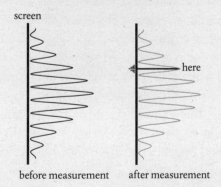

screen

here

before measurement   after measurement

Figure 7. Before measurement, the electron wave resulting from two-slit interference is presumed to form a pattern of alternating probability amplitude as shown on the left. After measurement, the electron is recorded to be 'here' at one, and only one, place on the screen. This is the 'collapse of the wavefunction'.

of 'explaining' how a distributed or delocalized quantum system becomes localized in a measurement. There is nothing in the mathematical prescription that forces (or even describes) the collapse.

Einstein didn't like this at all. Any such collapse would have to happen instantaneously, appearing to violate the principles of special relativity, which demands that no physical action or information having physical consequences can be transmitted faster than the speed of light. Indeed, it's not difficult to conceive of systems in which the collapse would somehow have to be 'transmitted' across vast distances. He called it *spooky action-at-a-distance*. Since Einstein had worked so hard to eliminate action-at-a-distance from the description of gravity, it's perhaps not difficult to understand why he might not like this particular feature of quantum theory.

Then there's the problem with the interpretation of quantum probability. The direct link between cause and effect has been an unassailable fact (and common experience) for all of human existence. In the world described by classical physics, if we do this, then that *will* happen. But in the world described by quantum physics, if we do this then that *might* happen, with a quantum probability that can only be derived from the specific pattern of peaks and troughs in the wavefunction. Einstein didn't like the notion of God playing dice with the universe, and arguments against this formed the basis for his famous debate with Bohr.

If we have no evidence that the collapse of the wavefunction actually occurs, then why assume that it happens only at the microscopic level of photons, electrons, and atoms? What if it happens only when we *look* to see where the electron has got to? This is the basis for the paradox of Schrödinger's cat, in which quantum probabilities are transferred up the measurement chain from atoms to a cat sitting in a closed box. The 'here' *and* 'there' nature of quantum probability translates into a cat which is both alive *and* dead. At least, until we open the lid of the box, and look. At which point we presume the wavefunction collapses and the fate of the cat is sealed.

Heisenberg's uncertainty principle places a fundamental limit on the precision with which we can measure the properties of quantum particles like electrons. These are properties such as the position of an electron in space and its momentum, or the energy of an electron and the rate of change of energy with time.

When he first formulated the principle, Heisenberg reasoned that on quantum scales it is simply impossible to make a measurement without *disturbing* the object under study in a wholly unpredictable way. No matter how precise or sophisticated our experimental techniques, we simply can't get past the fact that they will always be too 'clumsy', and this clumsiness is the source of the uncertainty. In this interpretation, such clumsiness prevents us from measuring the paths of each electron through a two-slit experiment (particles) *and* recording an interference pattern (waves) at the same time. The uncertainty principle places limits on what is *measureable*.*

But Bohr fundamentally disagreed, and the two argued bitterly.[9] Bohr reasoned that we reach for classical wave and particle concepts to describe the results of experiments because these are the only kinds of concepts accessible to us in our classical, macroscopic reality. Whatever the true nature of the electron, its behaviour is predicated by the kinds of measurements we choose to make. We conclude that in this measurement the electron is a wave. In another kind of measurement the electron is a particle. These measurements are mutually exclusive: we can ask questions

---

* It's surprising to hear how often this interpretation of the uncertainty principle is still taught by teachers who should really know better.

concerning the electron's wave-like properties and we can ask mutually exclusive questions concerning the electron's particle-like properties, but we cannot ask meaningfully what the electron *really is*.

Bohr concluded that these very different behaviours are not contradictory; they are instead *complementary*. When he first heard about the uncertainty principle from Heisenberg, he realized that this complementarity places a limit not on what is measureable, as Heisenberg had thought, but rather on what is *knowable*. This was absorbed into a general way of thinking about the meaning of quantum theory that became known as the *Copenhagen interpretation*.*

In the two-slit experiment, if we don't try to follow the paths of the electrons, then we get wave behaviour. But if we look to see *how* we get wave behaviour, we will get particle behaviour. These behaviours are complementary, and mutually exclusive. There is no way of tricking the electron to reveal both types of behaviour in the same experiment.

But the clumsiness argument that Heisenberg had used implies that the electron actually *does* possess precise properties of position and momentum, or follows precise paths and simultaneously exhibits interference. We could, in principle, measure these if only we had the wit to devise experiments that are more subtle and less clumsy. In contrast, Bohr believed that this has nothing to do with our ingenuity, or lack of it. It has everything to do with the nature of reality at the quantum level. We can't conceive an experiment of greater subtlety because such an experiment is simply *inconceivable*.

---

* By this time Bohr had established an Institute for Nuclear Physics in Copenhagen with support from the Carlsberg Foundation. The architects of the Copenhagen interpretation were Bohr, Heisenberg, and Pauli, though each differed in their views.

The problem runs very, very deep. Trying to eliminate or some-how deal with the collapse of the wavefunction and the notion of complementarity has spawned a veritable industry of variations and alternative interpretations of quantum theory. Perhaps there are indeed subtle mechanisms—collectively called 'hidden vari-ables'—which can account for wave-like behaviour in a system of particles, or particle behaviour in a system of waves. Such hidden variables may be local, meaning that they work at the level of individual particles. Or they could be non-local, for example involving some kind of 'pilot' wave guiding the individual particles along pre-determined paths. Not waves *or* particles, as complementarity demands, but waves *and* particles.

Einstein toyed with these ideas himself in 1927, but dismissed them as 'too cheap'.[10] Rather than attempt to solve these prob-lems, in his debate with Bohr he instead came up with a series of ever more ingenious thought experiments designed to show that quantum theory is either inconsistent or incomplete. Bohr stood firm. He resisted the challenges, each time ably defending the Copenhagen interpretation and in one instance using Einstein's own general theory of relativity against him.

But Bohr's defence relied increasingly on arguments based on clumsiness, on an essential and unavoidable disturbance of the system caused by the act of measurement, precisely of the kind for which he had criticized Heisenberg. To win the debate Einstein realized that he needed to find a challenge that did not depend directly on the kind of disturbance characteristic of a measure-ment, thereby completely undermining Bohr's defence.

Together with two young theorists Boris Podolsky and Nathan Rosen, in 1935 Einstein devised the ultimate challenge. The details are less important than the principles. The Einstein–Podolsky–Rosen (EPR) thought experiment is based on a physical system in which two particles are 'entangled', meaning that according to quantum theory their physics is governed by a single wavefunction. This wavefunction can be written as a combination (called a *superposition*) of all the different possible outcomes of a subsequent measurement.

As a consequence, the particles have properties that are *correlated*. It doesn't matter what kinds of properties these are: if one particle is measured to have one kind of property ('up', '+', 'vertical', and so on), then simple laws of conservation mean that the other particle *must* have properties that are correlated ('down', '−', 'horizontal', and so on). According to the Copenhagen interpretation, these two particles have no assigned properties until a measurement is made, and the wavefunction collapses. At this point some of the outcomes are realized—'up' and 'down' or '+' and '−'—with probabilities which reflect their weighting in the original wavefunction.

Now we see the real cunning in this challenge. Because of the way they have been created, we can determine the properties of one particle by making a measurement on the other. Now it doesn't matter how clumsy our measurement might be. Through the mechanism of entanglement we can discover— with certainty—the properties of a particle *without disturbing it in any way*. In practice we might be constrained by the dimensions of our laboratory, but we could, in principle, wait until one particle has travelled halfway across the universe before making our measurement on the other.

Does this mean that the collapse of the wavefunction must then somehow reach halfway across the universe—instantaneously— in order to 'fix' the properties of the distant particle? EPR wrote: 'No reasonable definition of reality could be expected to permit this.'[11]

Surely, it's easier to assume that these properties—whatever they may be—are fixed at the moment the particles are produced.

This one is 'up', that one is 'down', from the moment they are created. A measurement on one then simply tells us what property it had, all along, and by inference what property the other particle must have. But let's be absolutely clear. If we want to make this assumption, then we need to reach for an alternative local hidden variables theory.

Irish theorist John Bell realized that if such hidden variables really do exist, then in experiments with entangled particles the hidden variable theory will predict results incompatible with the predictions of quantum theory. It didn't matter that we couldn't be specific about precisely what these hidden variables were supposed to be. Assuming hidden variables of any kind means that the two particles are imagined to be *locally real*—they move apart as independent entities with defined properties and continue as independent entities until one, the other, or both are detected.

But quantum theory demands that the two particles are 'non-local', described by a single wavefunction. This contradiction is the basis for *Bell's theorem*.[12]

Bell published his ideas in 1966, and within a few years the first experiments designed to test Bell's theorem were carried out. The most widely known of these were performed by French physicist Alain Aspect and his colleagues in the early 1980s, based on the generation and detection of entangled photons. The results came down firmly in favour of quantum theory.[13]

Smolin was lucky. In the book *Albert Einstein: Philosopher-Scientist*, which he had drawn from his school library, there is a lively and very readable chapter by Bohr which details his famous debate with Einstein. With an appetite for the more philosophical or foundational aspects of quantum theory thus whetted, a teacher adopting an instrumentalist approach to the subject could easily have killed his interest. But in the spring semester of his first year in college Smolin was extraordinarily fortunate. He learned about quantum mechanics from Herbert Bernstein, by Smolin's own account a great physics teacher. The course concluded with detailed discussions of the EPR argument and Bell's theorem. 'Bell's paper was not yet widely known and had by that time very few citations,' Smolin explained. 'That was probably the first and only quantum mechanics course for undergraduates that included EPR and Bell.'[14]

When Smolin had read and come to terms with many of the original research papers, he retreated to his room and lay on his bed for a long time.

The EPR thought experiment pushed Bohr to drop the clumsiness defence, just as Einstein had intended. But this left Bohr with no alternative but to argue that the properties and behaviour of quantum particles must somehow be influenced by the kinds of measurements we make and how we *choose* to set up an apparatus an arbitrarily long distance away. This is a particularly uncomfortable notion. As English physicist Anthony Leggett put it many years later, 'nothing in our experience of physics indicates that the [experimental setup] is either more or less likely to affect the outcome of an experiment than, say, the position of the keys in the experimenter's pocket or the time shown by the clock on the wall'.[15]

Local hidden variable theories of the kind that Bell had considered are constrained by two important assumptions. In the first, we assume that due to the operation of the hidden variables, whatever measurement *result* we get for the first particle can in no way affect the result of any simultaneous or subsequent measurement we make on the second, distant particle. In the second, we assume that however we *set up* the apparatus to make the measurement on the first particle, this also can in no way affect the result we get for the second particle.

If we drop the second assumption but keep the first, the result is a kind of 'crypto' non-local hidden variables theory. The outcomes of measurements *are* affected by how we choose to set up our apparatus, just as Bohr claimed, but at least the results are in some sense preordained. In this extension of the theory, the wavefunction somehow 'senses' what's coming and is ready for it. This rids us of the collapse of the wavefunction though it does still leave us with some rather 'spooky' elements.

Leggett found that keeping the result assumption but dropping the setup assumption is still insufficient to reproduce all the predictions of quantum theory. Just as Bell had done in 1966, Leggett now derived the basis for a direct experimental test. This comes down to a rather simple question. Do quantum particles like electrons have the properties we assign to them (or indeed any properties) *before the act of measurement?*

These experiments were performed in 2006[16] and in 2010.[17] The results are pretty unequivocal. We must abandon both the result *and* the setup assumptions. Recent tests of Bell's inequality that close the door on various 'loopholes' were reported in 2015.[18] It seems that no matter how hard we try, or how unreasonable the resulting definition of reality, we just cannot avoid having to invoke the collapse of the wavefunction.

I don't want to mislead you. There are some ways out of this mess but I've always thought these alternative interpretations to be tinged with more than a hint of desperation, or even madness. Perhaps the collapse is indeed a real physical event, triggered when conscious beings choose to lift the lid and look. In this interpretation we conflate the deep mystery of quantum mechanics with the deeper mystery of consciousness, which, I confess, has never struck me as a particularly promising path to follow.

Or perhaps the collapse doesn't occur at all, and instead the entire universe splits into parallel versions of itself. In one of these 'many worlds', the electron is 'here', the cat is alive, and Gwyneth

Paltrow boards the train, eventually to die in the arms of her new love.* In other worlds the electron is 'there', the cat is dead, and Gwyneth Paltrow misses the train, and lives to tell the tale.

Is there a better way? Perhaps a theory of quantum gravity can help to shed some light.

---

* The Gwyneth Paltrow reference is taken from the plot of the 1998 romantic drama *Sliding Doors*, directed by Peter Howitt.

4

MASS AIN'T WHAT
IT USED TO BE

When I first saw the sequence of photographs showing the emergence of an interference pattern as more and more electrons are passed—one at a time—through two slits (Figure 6; see Chapter 3), I felt really rather uneasy. I had studied quantum mechanics and understood the principles well enough, but I couldn't quite let go of my naïve classical preconceptions. If an electron wave passes through both slits simultaneously and interferes with itself, to be eventually detected as a single dot on a distant screen or photographic plate, what happens to the electron's *mass* in between?

Getting to an answer for this question takes us on the journey that leads to a truly marvellous quantum theoretical structure called the *standard model of particle physics*.

The version of quantum mechanics developed by the pioneer theorists of the 1920s and early 1930s was revolutionary and, as we've seen, really quite extraordinary in its implications for our understanding of matter and radiation. It remains perfectly valid and relevant today (and it is still taught to science students pretty much in its original form) but it is also quite limited. It can be used to describe quantum systems involving particles which maintain their integrity in physical processes, such as an electron moving between different orbits in an atom or through a two-slit experiment. But it can't be used to describe situations in which particles are created or destroyed, and so it excludes quite a lot of interesting physics.

Physicists such as Heisenberg, Pauli, and Dirac realized that they needed a theory of *quantum fields*. Such a field has a strength or intensity at every point in space, and so is delocalized much like an extended, three-dimensional wave. In a classical wave theory, energy can be added or taken away from a field continuously. In contrast, energy is added or taken away from a quantum field only in discrete quanta.

In his *Lectures on Quantum Mechanics*, the published version of a series of lectures delivered at Yeshiva University in New York City in 1964, Dirac summarized it like this: 'according to the general ideas of de Broglie and Schrödinger every particle is associated with waves and these waves may be considered as a field.'[1]

The quanta themselves are then reinterpreted as characteristic fluctuations, disturbances, or excitations of the quantum field. In quantum field theory, wave–particle duality is expressed in terms of localized fluctuations (particles) of the spatially extended, non-local field (waves). We are then perfectly at liberty to talk about the properties and behaviour of the field itself, or the properties and behaviour of its quanta. These two representations are equally valid (Bohr would say they are complementary), but as far as measurement is concerned, they are mutually exclusive.

It seemed clear that the photon must be the quantum of the electromagnetic field, created and destroyed when electrically charged particles interact. A proper quantum field theory should then describe the interactions between an electron field (whose quanta are electrons) and the electromagnetic field (whose quanta are photons).

The obvious first step was to find a way to introduce quantum-like properties and behaviour into Maxwell's classical theory of the electromagnetic field. It was understood that if this could be done in a way that satisfied the demands of special relativity the result would be a quantum version of electrodynamics, called *quantum electrodynamics* or QED.

Heisenberg and Pauli developed an early version of QED in 1929, but ran into some big problems that were not resolved until 1947. The solution involves some mathematical sleight-of-hand, of a kind that signals the need for a radical rethink of the very concept of mass.

Now, to be totally fair, even in classical physics the concept of mass is more mysteriously elusive that we would ordinarily like to admit. The definition of mass that Newton provided in the *Mathematical Principles* is deliciously circular, and so doesn't define mass at all. The arch-empiricist Mach tried to clear all this up by developing an operational definition, essentially making mass—whatever it is—relative to a 'standard'.

Of course, none of this shakes our confidence, sprung from a lifetime's experience of the physical world, that we at least know what mass *does*.

To understand what was needed to solve the problems of QED we need two essential insights from the quantum world. The first concerns the way in which, in a quantum field theory, forces are *transmitted* between fields or particles. We distinguish between elementary particles that govern the properties and behaviour of matter—such as electrons—and elementary particles (such as photons) that transmit or 'carry' forces between the matter particles. This distinction was first articulated by German physicist Hans Bethe and Italian Enrico Fermi in 1932.

Elementary particles are distinguished not only by their electric charge and mass, but also by a further property which we call *spin*. This choice of name is a little unfortunate, and arises because some physicists in the 1920s suspected that an electron behaves rather like a little ball of charged matter, rotating on an axis much like the Earth rotates on its axis as it orbits the Sun. This is not what happens, but the name stuck.

Spin is indeed a measure of a particle's intrinsic *angular momentum*, the momentum we associate with rotational motion. But I'm afraid this is not as simple as a particle spinning around an axis. Put it this way, if we really wanted to push this analogy, then we would need to accept that an electron must spin *twice* around its axis to get back to where it started.[2] This is reflected in the fact that the electron has a spin quantum number of ½ and consequently has only two spin orientations—only two directions it can 'point' in a magnetic field—which we call 'spin up' (+½) and 'spin down' (−½).

The spin properties of the electron place it in a class of particle known as *fermions* (named for Enrico Fermi). It turns out that all matter particles are fermions. Force particles, in contrast, are *bosons* (named for Indian physicist Satyendra Nath Bose), which means that their spin quantum numbers take integral values. The spin quantum number of the photon is 1.

Armed with this understanding, we can now picture what happens when two electrons interact. We know that like charges repel—so as the electrons approach each other we imagine they 'feel' a force of electrostatic repulsion, which then hurls them apart. Moving electrons generate an electromagnetic field (this is the basis for all radio communications), and in quantum field theory the interaction is mediated by the electromagnetic field and represented as the *exchange* of a photon. This changes the electrons' relative speed and direction, and they move away from each other. We don't 'see' the photon that passes between them, and so this is called a 'virtual' photon, on the understanding that being virtual doesn't mean it's not 'real'.

Richard Feynman developed a simple visual method for keeping track of this kind of interaction based on simple line drawings which are now called *Feynman diagrams*. In a Feynman diagram we map the motions of the particles to a generalized 'space' (representing three-dimensional space) and 'time'. The particle motions are drawn as straight lines and the force they experience is represented as a virtual force particle which passes between them, drawn as a wavy line (see Figure 8, which shows a generalized repulsive force acting on two matter particles).

The second insight concerns Heisenberg's uncertainty principle. It is easy to fall into the trap of thinking that this principle is really rather constraining, placing as it does limits on the

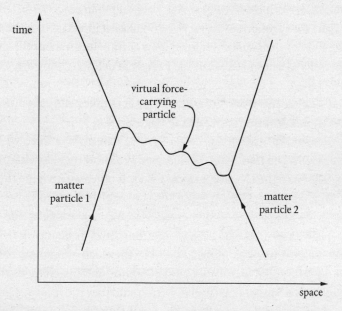

Figure 8. In a Feynman diagram, the motions of matter particles are represented by straight lines drawn through 'space' and 'time'. Forces are transmitted between them by force particles, shown as wavy lines.

precision with which we can measure certain combinations of properties, such as position and momentum or energy and time. But the principle cuts both ways.

Suppose we were somehow able to create a complete vacuum, in which the energy of some quantum field (such as a quantum electromagnetic field) is zero. With nothing at all in this vacuum able to exert any physical change, we would quickly conclude that the rate of change of the energy of this field with time must also be zero. But then we fall foul of the uncertainty principle. Zero is a pretty precise measure, so in a vacuum the energy of a quantum field and its rate of change with time *cannot* both be precisely zero.

The uncertainty principle doesn't expressly forbid 'borrowing' the energy required to create virtual photons or even virtual matter particles literally out of nothing, so long as this energy is 'given back' within a time that conforms to the demands of the principle. The larger the energy borrowed, the sooner it must be given back.

The vacuum is therefore filled with random quantum fluctuations, like turbulent waves in a restless ocean. But we know that fluctuations in a quantum field are equivalent to particles. These are virtual particles, like a kind of background 'noise', which average out to zero, in terms of both energy and its rate of change. But they can nevertheless be non-zero at individual points in space-time. 'Empty' space, is in fact, a chaos of wildly fluctuating quantum fields and virtual particles.* Occasionally, these random fluctuations strike it lucky, and out pops an elementary particle or a particle paired with its antiparticle (its antimatter partner), which then mutually annihilate.

---

* Experimental evidence for the *Casimir effect*, named for Dutch theorist Hendrik Casimir, provides some empirical proof that this does indeed happen.

Now we bring these two insights together. An interaction involving the electromagnetic force doesn't just involve the exchange of a virtual photon. It may also involve two virtual photons, or a virtual photon which spontaneously creates an electron and antielectron pair which then mutually annihilate to form another photon. The possibilities are actually endless, depending only on what's allowed within the constraints of the uncertainty principle. But the more elaborate and convoluted the process, the lower its probability and the smaller its contribution to the overall interaction. These small contributions are called *radiative corrections*, and physicists make extensive use of Feynman diagrams to take them into account.

The theorists realized that the problems with the early versions of QED were a result of the electron interacting with its own self-generated electromagnetic field, causing some terms in the equations to mushroom to infinity. As a result of these interactions, the electron gathers a covering of virtual particles around itself. These virtual particles have an energy and, as we know from $m = E/c^2$, the mass of such a 'dressed' electron is therefore greater than its 'bare' mass, or the mass the electron would be expected to possess if it could be separated from its own electromagnetic field.

It's impossible to know what the bare mass of the electron is, but the equations of QED could now be manipulated to solve the problems. The theorists discovered that subtracting the equation describing the electron in one physical situation from the equation describing the electron in a different situation meant they could get rid of the infinite terms. Subtracting infinity from infinity doesn't seem on the surface to be a very sensible thing to attempt, but it was found that the result was not only finite, it was also right. This sleight-of-hand is called *mass renormalization*. Dirac, the mathematical purist, always thought it was 'ugly'.

But there could be no denying the power of a fully relativistic version of QED, one that conforms to the demands of Einstein's special theory of relativity. The g-factor of the electron is a physical constant which governs the interaction between an electron and a magnet. It is predicted by QED to have the value 2.00231930476. The comparable experimental value is 2.00231930482.

'To give you a feeling for the accuracy of these numbers,' wrote Feynman, 'it comes out to something like this: If you were to measure the distance from Los Angeles to New York to this accuracy, it would be exact to the thickness of a human hair.'[3]

This extraordinary success led physicists in the 1950s to seek out quantum field theories that could describe the other forces of nature. It was understood that electromagnetism is the force that holds together positively charged atomic nuclei and negatively charged electrons inside atoms. But observations and experiments had revealed the existence of a further two forces at work *inside* the nucleus itself.

At this time it was thought that there were just two kinds of matter particles to be found in the nucleus—positively charged protons and electrically neutral neutrons, each with similar (though not identical) masses about 2000 times heavier than the electron. The force binding protons and neutrons together in the nucleus was called the *strong nuclear force*.

We might be tempted to think that the strong nuclear force and electromagnetism ought to be enough, but physicists discovered that the neutron is rather unstable. It is susceptible to radioactive decay, in the process transforming into a proton and ejecting a high-speed electron and a rather odd particle called a *neutrino* (Italian for 'little neutral one').* This process could only

* Actually, in the radioactive decay of a neutron, the neutron transforms into a proton and a high-speed electron and *antineutrino* are emitted.

be explained by invoking the existence of a third force, which was called the *weak nuclear force* or *weak interaction*.

The theorists soon hit more problems. Although the strength of the electromagnetic force falls off with distance, it is a long-range force that can in principle stretch off to infinity. This kind of force can be carried quite happily by massless photons moving at the speed of light. But the strong and weak nuclear forces had to be very short-range forces—they are confined to work *within* the nucleus.

In 1935 Japanese physicist Hideki Yukawa suggested that, in contrast to the massless photons of electromagnetism, the carriers of short-range forces ought to be 'heavy' particles. He estimated that the carriers of the strong nuclear force binding protons and neutrons together inside the nucleus should have masses about 200 times heavier than the electron. Such force carriers could be expected to move rather sluggishly between the matter particles, at speeds much less than light.[4]

But, try as they might, in the early 1950s theorists just could not formulate a quantum field theory that could accommodate heavy force carriers. Something was missing.

There was a clue from an almost forgotten, rather outrageous piece of speculation that had been published in 1941 by American theorist Julian Schwinger (who would go on to become one of the architects of QED). He noticed that if the weak force was assumed to be carried by a massless particle—like the photon—then it would actually have a strength and range entirely equivalent to electromagnetism. To all intents and purposes, it looked as though the weak force and electromagnetism had once belonged to one and the same unified 'electro-weak' force.

Schwinger passed this challenge on to his Harvard graduate student, Sheldon Glashow. After a few false starts, Glashow developed a quantum field theory of weak interactions in which the weak force is carried by three particles. Two of these carry electrical charge and are now called the W^+ and W^- particles. This leaves a third, neutral force carrier now called the Z^0. But the theory said that these particles should all be massless, just like the photon, making the weak force long range rather than short range. If Glashow tried to fudge the equations by adding masses 'by hand', then the theory couldn't be renormalized.

Something must have happened to the carriers of the weak force. They had somehow 'gained mass', thereby splitting the electro-weak force into the two distinct forces we recognize today. Physicists have a way of thinking about this kind of thing that has deep mathematical as well as physical significance. They call it 'breaking the symmetry'. Whatever had happened had broken the symmetry of the electro-weak force, creating two new forces.

The details were worked out in a number of research papers published in 1964.* In a sense, there was nothing in the quantum field theories developed in the 1950s that could break the symmetry of the electro-weak force. There was nothing for the force-carrying particles to 'hang on to', nothing to slow them down in a way that we then *interpret* as equivalent to gaining mass. What was needed was another, distinctly different, kind of quantum field that the particles could interact with, and hang on to.

Physicists trying to explain what's going on here have used various analogies, the most popular suggesting that this mysterious new quantum field behaves rather like molasses, dragging on the particles and slowing them down, their resistance to acceleration manifesting itself as mass. Such analogies are

---

* I don't want to confuse you unduly, but the objective of these 1964 papers was actually a quantum field theory of the strong nuclear force. It so happens that the principles are much the same, and—as we will soon see—the solution they worked out was applied to break the symmetry of the electro-weak force just three years later.

always inadequate, but at least they help us try to get our heads around what's happening.

From about 1972 this mysterious new quantum field has been known as the *Higgs field*, named for English theorist Peter Higgs, author of one of the papers published in 1964. The characteristic quantum fluctuation of the Higgs field is a particle called the *Higgs boson*.

The development of this mechanism opened the door to a fully fledged quantum field theory of the weak force, published by American theorist Steven Weinberg in 1967 and Pakistan-born theorist Abdus Salam in 1968.* The last remaining obstacle was removed in 1971 when Dutch theorists Martinus Veltman and Gerard 't Hooft showed that this theory can be renormalized.

In 1967 Weinberg had used the theory to predict the masses of the W and Z particles, which we can think of as 'heavy' photons. They were discovered in particle collider experiments at CERN, the European Organization for Nuclear Research in Switzerland, in 1982 and 1983, with masses very close to Weinberg's prediction.

Adding a background Higgs field suggests that, whatever it is, this field pervades the entire universe like a modern-day ether. Although the Higgs mechanism is used to explain how the W and Z particles gain mass, it turns out that this is how *all* particles gain mass, including matter particles such as electrons.

In the absence of the Higgs field, all particles would be massless, travelling at the speed of light. There would be no mass and no material substance. There would be no universe of stars and galaxies, no planets, no life, and no *Homo sapiens*. This is one of

* Salam actually held back from publishing his results as he wanted to extend the theory to include baryons and mesons (Weinberg had considered only leptons). But this proved impossible, and the only record of Salam's efforts in this period are the published proceedings of the Eighth Nobel Symposium held in Gothenburg, Sweden, on 19–25 May 1968.

the reasons why the American particle physicist Leon Lederman decided to call the characteristic particle of the Higgs field the 'God particle'.[5] Most physicists hate this name (Carlo Rovelli would later claim that the name 'is so stupid as to be unworthy of comment').[6] But there can be no doubt that it caught the attention of the public.

We had to wait almost another thirty years for confirmation that the theory is essentially correct. The Higgs boson was first discovered at CERN in July 2012.

This takes care of the weak force and electromagnetism, but what about the strong nuclear force? This brings us to one of the most extraordinary parts of the story. To make some sense of the seemingly bizarre pattern in the properties of the 'zoo' of particles that had been discovered by the early 1960s, American theorists Murray Gell-Mann and George Zweig had independently suggested that protons and neutrons might not, after all, be elementary. The pattern suggested that protons and neutrons might each consist of three even more elementary particles which Gell-Mann referred to as *quarks*.*

This might sound quite reasonable, but to account for the simple fact that protons are positively charged (with charge +1) and neutrons are neutral (with charge 0) it was necessary further to suggest that the quarks carry *fractional* electric charges. To make a proton we need two quarks with a charge of $+\frac{2}{3}$ and one quark carrying $-\frac{1}{3}$ to get an overall charge of +1. One quark with a charge of $+\frac{2}{3}$ and two carrying $-\frac{1}{3}$ would then give a neutron with overall zero charge.

The positively charged quarks became known as 'up' quarks, and their negatively charged companions as 'down' quarks. To explain the pattern of other particles known at the time a third

* Zweig called them 'aces'.

quark was required: essentially a heavier version of the down quark which was called the 'strange' quark.

The terms up, down, and strange are called quark *flavours*. It was now understood that a weak force interaction changes the flavour of one of the quarks inside a neutron, transforming a down quark into an up quark and ejecting an electron and an antineutrino. This turns the neutron into a proton.

Nobody really knew what to make of this, and most physicists were sceptical. When Weinberg sat down in 1967 to formulate a quantum field theory using the Higgs mechanism, he deliberately avoided involving quarks in the picture because he 'simply did not believe in quarks'.[7]

This situation began to change in 1968, when experiments at the Stanford Linear Accelerator Center in California revealed hints that protons and neutrons are indeed composite particles. But there were more puzzles. These experiments suggested that, far from being held tightly inside the proton, as we might anticipate, the quarks actually rattle around as though they are almost entirely free. Now, to this day, nobody has ever seen a free quark in a particle collider experiment. If these are supposed to be elementary building blocks, which rattle around rather loosely inside protons and neutrons, why don't they ever come out?

Our familiarity with Newtonian gravity and electromagnetism leads us to think about a force of nature as something that is centred on a point—typically, the centre of the particle or object which 'generates' the force, and which then declines in strength the further away we get from it. The pull we feel between the north and south poles of two bar magnets gets weaker as we move them apart.

But the force holding quarks together inside protons and neutrons doesn't behave this way at all. In 1973, Princeton theorists David Gross and Frank Wilczek and Harvard theorist David

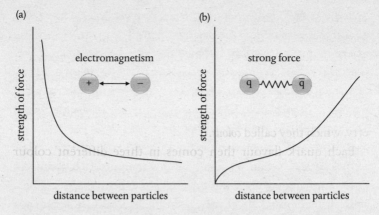

Figure 9. The electromagnetic force between two electrically charged particles increases in strength as the particles move closer together (a). But the force that binds quarks together behaves differently (b). In the limit of zero separation between a quark and an anti-quark (for example), the force falls to zero. The force increases as the quark and anti-quark are pulled apart.

Politzer showed that this force instead acts as though adjacent quarks are fastened together by a piece of strong elastic or a spring.* As the quarks are pushed close together, the elastic or the spring becomes loose, and the force between them diminishes (see Figure 9). Inside a proton or neutron the quarks are fastened together but are close enough to 'rattle around'. But if we try to pull the quarks apart, the force 'kicks in', as the elastic or the spring gets taut.

It turns out that the energies required to break the elastic or spring that holds two quarks together is more than enough to allow other quarks to be conjured from the vacuum to fill any

---

* Gerard 't Hooft arrived at this same conclusion a little earlier, and mentioned it at a small conference in Marseille in 1972. His colleague Kurt Symanzik urged him to publish the result: 'If you don't someone else will.' 't Hooft did not heed his colleague's sensible advice.[8]

vacancies. Quarks just do not like to be separated, which is (presumably) why they've never been seen in isolation.

This kind of force is very different from electromagnetism or the weak force, so it can't be traced to the quarks' electric charges or their flavours. Theorists Gell-Mann and Harald Fritzsch had no alternative but to propose that quarks possess a further property, which they called *colour*.

Each quark flavour then comes in three different colour varieties—red, green, and blue—and protons and neutrons are formed from three quarks of each colour. For example, a proton might be formed from a red up quark, a green up quark, and a blue down quark. Such a scheme demands a system of eight coloured, massless force-carriers which the theorists called *gluons*, which behave much like a piece of elastic or a spring, and they started to refer to the strong force between quarks as the *colour force*.

With these ingredients, in 1973 Gell-Mann, Fritzsch, and Swiss theorist Heinrich Leutwyler were able to craft a quantum field theory of the colour force. The result is quantum chromodynamics (QCD).

Further particle discoveries have added heavier versions of the up quark (called charm and top) and a heavier version of the down and strange quarks (called bottom). A heavier version of the familiar electron was discovered in 1936. This 'heavy electron' is now called the muon, and an even heavier version (called the tau) was discovered in the mid-late 1970s. These are partnered with a family of neutrinos—the electron neutrino, muon neutrino, and tau neutrino. The electron, muon, tau, and their neutrinos are collectively called *leptons*.

In the standard model of particle physics, the matter particles are organized in three 'generations' of quarks and leptons, with the first generation of up and down quarks, the electron and electron neutrino accounting for the atoms of ordinary material

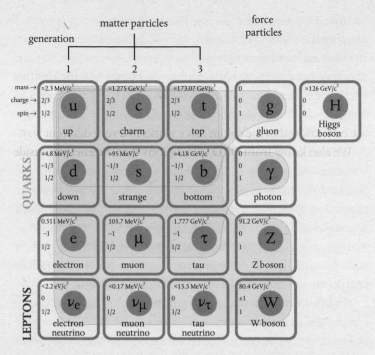

Figure 10. The standard model of particle physics describes the interactions of three generations of matter particles through three kinds of force, mediated by a collection of 'force carriers'. The masses of the matter and force particles are determined by their interactions with the Higgs field.

substance and their interactions. There's nothing in principle preventing the discovery of members of a fourth generation, although there's some experimental evidence and compelling theoretical arguments to suggest that three generations are all there is.

To the matter particles we must add the particles responsible for carrying forces between them. These are the massless photon (electromagnetic force), the W and Z particles (weak force) and eight massless, coloured gluons (colour force). The Higgs boson then completes the standard model, summarized in Figure 10.

Now, I haven't forgotten Yukawa's prediction, mentioned earlier, that the carriers of the strong nuclear force acting between protons and neutrons should be heavy particles with about 200 times the mass of the electron. Surely, this contradicts the idea that the colour force is carried by massless gluons? But Yukawa was thinking of a force centred on a point, like electromagnetism, and we now know that the colour force works in a very different way.

We also know from QCD calculations that the energies inside protons and neutrons are so high that each quark produces a blizzard of virtual gluons that pass back and forth between them, together with quark–antiquark pairs. Physicists sometimes call the three quarks that make up a proton or a neutron 'valence' quarks, as there's enough energy inside these particles for a further 'sea' of quark–antiquark pairs to form. The valence quarks are not the only quarks inside these particles.

And this is the answer. Although the colour force acts on the quarks and gluons inside protons and neutrons, it 'leaks' beyond the confines of these particles and makes them 'sticky'. This leakage takes the form of quark–antiquark pairs called *pions*, which come in positive, negative, and neutral varieties. We can then think of the pions as carriers of a residual or secondary force which then acts between protons and neutrons. The pions have masses around 265 and 274 times the mass of the electron, so Yukawa's prediction was pretty good.

This brings us rather neatly to our final reflection on the nature of mass. We've already seen that all matter and force particles which possess mass gain this through their interactions with the Higgs field. We've also seen that the masses of matter particles like electrons depend on the energy of the covering of virtual particles that gather around and 'dress' it.

Since as much as 99 per cent of the mass of an individual atom resides in the protons and neutrons in its nucleus, we might be tempted to think we should be able to track this back to the masses of the individual up and down quarks that compose them. But although estimating the masses of the up and down quarks is a bit of a messy business, with many assumptions and approximations, we can nevertheless be very confident that that total mass of three quarks falls substantially short of the measured masses of the proton and neutron.

In fact, elaborate QCD calculations demonstrate that about 95 per cent of the mass of a proton or neutron is derived from the *energy of the massless gluons* that carry the colour force between the quarks.

We see that Einstein's great insight was indeed that $m = E/c^2$— the mass of a body is a measure of its energy content—and we conclude that mass is not something that an object *has* (it is not a property). Instead, the mass of an object is something that it *does* (it is rather a behaviour).

The electron field passes at once through both slits, its pattern of quantum probability shaped by the interference in the field. The electron, interpreted as a characteristic fluctuation of the field, is then manifested only when the field interacts with the screen or photographic plate. At this moment the wavefunction 'collapses' and what we recognize as the electron—together with its spin, charge, and mass—is plucked as if from thin air.

5

HOW TO FUDGE THE EQUATIONS OF THE UNIVERSE

Physics is not a subject that appeals to everybody. Even those students sufficiently motivated to commit three or four years of their lives to study for a degree in physics will tell you that there are some parts of the subject that are, frankly, a bit dull. But for those drawn to physics by their fascination with deep questions concerning physical reality, for those bearing the burden of a profound and irresistible longing to understand the secrets of nature, there are many compensations. First, grasp the intricate details of Einstein's general theory of relativity and absorb a few of the lessons of quantum mechanics, and in the next class you get to learn how to apply it all to the entire universe.

Awesome.

Theories of the entire universe belong in a subcategory of physics called *physical cosmology*. Of course, philosophers have speculated on the nature of the universe ever since the ancient Greeks, devising ever more elaborate systems of 'celestial mechanics'. Newton had demonstrated that his laws of motion and universal gravitation apply to the planets of the Solar System. But, even by the standards of twentieth-century science, extending physical theories to describe the entire universe was rather novel, and regarded with deep suspicion by many in the physics community. Cosmology was not generally considered to be a 'proper' science until the 1960s.

Einstein understood what he was getting himself into. In 1917, shortly before presenting another paper to the Prussian Academy of Sciences on the cosmological implications of general relativity, he wrote to his friend and colleague, Austrian theorist Paul Ehrenfest: 'I have...again perpetrated something about gravitation theory which somewhat exposes me to the danger of being confined in a madhouse.'[1]

Two years after presenting his gravitational field equations, Einstein was still in pursuit of a physical interpretation of Mach's principle. If the water flows up the inside of Newton's bucket because the entire universe is spinning around it, then the origin of inertia had to be traced back to the content and large-scale behaviour of the universe itself.

He first had to decide what kind of universe to which to apply his field equations. Newton's universe is necessarily spatially infinite. This is a problem because if an object's inertia is meant to be governed by the cumulative action of all the other objects in the universe, then an infinite number of objects implies an infinite inertia. Applying Mach's principle in Newton's infinite universe means that nothing ever moves.

But a finite universe brings other problems, not least the logic that suggests that it ought to have 'edges', boundaries where the universe comes to an end (and begging awkward questions about what might lie beyond). To avoid this problem, it is necessary to conceive of a universe which is spatially finite, but which has no edge. In other words, we need to imagine a universe which is finite, but unbounded. At first sight this seems logically impossible, but only if the universe is imagined to be flat. By now

Einstein was well versed in the geometries of things that curve. We know well enough that the Earth is finite—it has a mass that can be estimated (it's about six million billion billion kilograms). And, although from any human vantage point on its surface the Earth looks flat, we know that it is spherical, and so has no edge. The Earth is finite but unbounded.

So, in Einstein's universe spacetime curves back on itself like the surface of a sphere, and contains a finite amount of mass-energy. Note that the 'sphere' in question is a sphere in four space-time dimensions (what mathematicians refer to as a 3-sphere or, more generally, a hypersphere—see Figure 11). The surface of

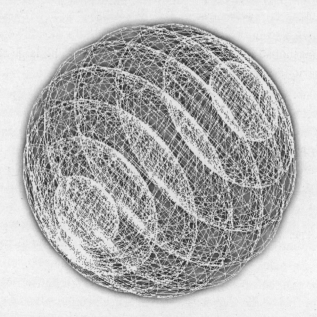

Figure 11. This representation of a 3-sphere (or hypersphere) is a projection into three-dimensional space, showing the structure as a stack of regular spheres (2-spheres), which is then projected onto two dimensions.

the Earth, in contrast, is a 2-sphere, a sphere in three-dimensional space. If it were possible for us to set off together across Einstein's universe on a journey along a straight path, we would eventually find ourselves returning to our point of departure.

Curiously, this was not the first example of a finite, but unbounded universe. Carlo Rovelli was delighted to discover that Einstein had been anticipated by the great Italian poet Dante Alighieri, who had studied at the University of Bologna in the fourteenth century. In *Paradiso*, the third and final part of the *Divine Comedy*, Dante describes his ascent to the Empyrean, the abode of God and the angels. Confronted with the challenge of describing the universe at its very edge, he fixes on the same solution that Einstein would choose, five hundred years later: 'Dante embeds the model in four dimensions, which does, as we know, solve the problem.'[2]

Einstein now had to deal with a more stubborn second problem. Gravity pulls objects together but it doesn't push them apart. Unlike electricity or magnetism, gravity doesn't have opposite charges or opposite poles, which attract *and* repel. Put another way, mass-energy curves spacetime only in ways which cause objects to move towards each other, never away.

Newton well understood where this leads. The mutual gravitational attraction between all the objects in an infinite universe would cause the universe to collapse in on itself. He had no choice but to invoke the ultimate solution, arguing in the *Mathematical Principles* that God had placed the stars sufficiently far apart to prevent such a catastrophic collapse from happening.[3]

Several centuries of astronomy had certainly produced no evidence that all the stars in the universe are rushing towards each other and in 1917 plain common sense suggested that the universe is surely *static*, not dynamic—we don't call them 'fixed stars' for nothing. But, just as in Newton's gravity, there was nothing in Einstein's field equations that would help to hold the universe steady. Einstein needed a physical mechanism that would do much the same job that Newton had asked of God.

His solution was to introduce a rather unsatisfactory 'fudge'. The left-hand side of the gravitational field equations describes the extent of spacetime curvature that will determine the motions of all the mass-energy, which is summarized on the right-hand side. Despite their extraordinary success, Einstein concluded that his equations were still 'unbalanced'. To engineer a static universe, he attempted to balance the equations by introducing a new term on the left-hand side, imbuing spacetime with a curious kind of antigravitational force, a negative gravitational pressure which builds in strength over long distances, counteracting the effects of the curvature caused by all the mass-energy on the right-hand side.

The new term is now characterized by a parameter Λ (Greek lambda), which is referred to as the 'cosmological constant'. By carefully selecting the value of Λ, Einstein found that he could achieve perfect balance: a static universe. He admitted that he 'had to introduce an extension to the field equations that is not justified by our actual knowledge of gravitation'.[4] But at the time it must have seemed like a neat solution. The cosmological constant didn't alter the way general relativity works over short distances, so all the theory's successful predictions were preserved.

However, Einstein had also made some significant errors. Not only was there no evidence or even a real theoretical justification for the cosmological constant, it actually doesn't do what Einstein claimed. The universe that he described in his 1917 paper is actually not static and is instead rather unstable. The modified field equations still didn't completely exclude the possibility that the universe might contract or expand.

Fourteen years later, Einstein's universe was overtaken by events.

In 1917, the visible universe was thought to consist of just the few hundred billion stars in our own Milky Way galaxy, dotted here and there with rather diffuse objects, called *nebulae*. For some years the American astronomer Vesto Slipher, working at the Lowell Observatory in Flagstaff, Arizona, had been making use of the Doppler Effect to investigate the relative speeds of these nebulae (see Figure 12).* As he gathered his data, he found that most are moving away from us, with speeds up to an astonishing 1,100 kilometres per second.

Because they are moving away, light from the nebulae is shifted to lower frequencies or longer wavelengths (the light waves are stretched out). In the visible spectrum, frequencies increase from red to violet, so a shift to lower frequencies is equivalent to a shift

blueshift

redshift

Figure 12. Suppose a stationary object emits a sound or light wave with a wavelength λ. If the object is set in motion, the wave emitted in the direction of travel becomes compressed—there are more cycles per unit distance or the wave has a shorter wavelength. The wave emitted in the opposite direction becomes stretched out—there are fewer cycles per unit distance or the wave has a longer wavelength. This is the Doppler Effect.

---

* When a moving object emits a wave signal (light or sound), the pitch (or frequency) that we detect depends on whether it is moving towards or away from us, and the speed with which it is travelling. The Doppler Effect is familiar to anyone who has listened to the siren of an ambulance or police car as it speeds past.

towards the red end of the spectrum. This is why the shift is called a *redshift*.

Arguments developed that the nebulae are complete galaxies in their own right, lying far outside the Milky Way. The matter was resolved in December 1924, when American astronomer Edwin Hubble estimated that a nebula found in the constellation of Andromeda is just short of a million light-years distant.* The Great Andromeda Nebula was quickly renamed the Andromeda *Galaxy*, and overnight the visible universe got an awful lot bigger.

Working together with his assistant Milton Humason at the Mount Wilson Observatory near Pasadena, California, Hubble subsequently extended Slipher's data set to include the redshifts of many more galaxies and established a surprisingly simple relationship between speed and distance, which is known as Hubble's law.[5] The more distant the galaxy, the faster it is moving away. There could really be only one explanation. The universe is not static at all; it is *expanding*.

That most of the galaxies are moving away from us does not place us at the centre of the universe. In an expanding universe it is spacetime that is expanding, with every point in spacetime moving further away from every other point (see Figure 13). The redshift in the frequency of light emitted by distant galaxies is actually *not* caused by the Doppler Effect, after all, as this involves the motion of light waves *through* space. It is instead caused by the expansion of spacetime in which the light is travelling towards us, and is called a *cosmological redshift*.

Accepting the evidence of an expanding universe meant accepting what this implies for its evolutionary history. Simple

---

* A light-year is the distance that light travels in a year, about 9,500 billion kilometres. The diameter of the Milky Way is estimated to be between 100,000 to 120,000 light-years, so the Andromeda nebula is well outside this.

(a)

(b)

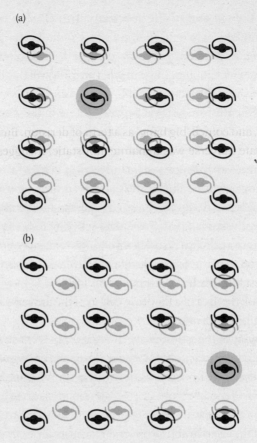

Figure 13. Imagine a universe consisting of a uniform distribution of galaxies. The pattern in grey represents this universe at some moment in time. The pattern in black is the distribution some time later, after the universe has expanded. From the perspective of the galaxy in row 2, column 2, shown in (a), all the galaxies around it have moved away, with more distant galaxies appearing to have moved the furthest. In (b), we fix on the galaxy in row 3, column 4. The result is the same—all the galaxies have moved away, with more distant galaxies moving the furthest. The fact that we see most galaxies receding from us doesn't mean we're at the centre of the universe.

extrapolation backwards in time suggests that there must have been a moment when all the mass-energy in the universe was compacted to an infinitesimally small point (called a 'singularity'), from which it burst forth in what we now call the 'Big Bang'.

Not everyone was comfortable with this idea. The maverick English physicist Fred Hoyle argued instead for a 'steady-state' universe, and coined 'big bang' as a term of derision. But even the steady-state universe was dynamic, not static. It suggested that, as spacetime expands, new matter is continuously created so that the large-scale observable universe is unchanged.

Either way, the cosmological constant, introduced in an attempt to engineer a static universe, was deemed to be unnecessary, and even a bit of an embarrassment. Einstein later came to regret this attempt to fudge the equations of the universe, suggesting it was the biggest blunder he ever made in his life.[6]

Few took the idea of a Big Bang origin of the universe seriously. But Ukrainian-born émigré physicist George Gamow realized that if this was indeed how the universe was born, then there ought to be consequences that are observable today. Working with his American postgraduate student Ralph Alpher, in 1948 Gamow modeled the nuclear physics of a 'primordial' universe consisting of protons, neutrons, and electrons, and successfully predicted the relative abundance of hydrogen and helium atoms.

The physics suggested that, as the early universe expanded and cooled, protons and neutrons would first react together to form small atomic nuclei, then these nuclei would join with free electrons to form neutral hydrogen and helium atoms, in a process called 'recombination'.

Up to this moment, the universe was a 'plasma' of electrically charged particles, with electromagnetic radiation dancing back and forth between them forming an impenetrable fog. But after recombination, the radiation would have nowhere else to go. It would have been released, clearing the fog, making the universe transparent and bathing it in a kind of background 'glow'. Gamow forged ahead, and later in 1948 he submitted a paper predicting the densities of matter and radiation at this moment of recombination.

But his calculations were quite wrong. Alpher, and another colleague Robert Herman, put them right and went on to predict that the radiation released by recombination would have persisted to the present day, and would pervade the entire universe. They estimated that this cosmic background radiation would be characterized by a temperature just a few degrees above zero on the absolute temperature scale.* At this temperature it would have a distribution of frequencies which peaks in the microwave and infrared.

The cosmic background radiation was famously detected in 1964—purely by accident—by radio astronomers Arno Penzias and Robert Wilson working at the Bell Laboratories' Holmdel research facility in New Jersey. They had been looking for microwave radiation that was thought to be emitted by the glowing cloud of gas surrounding our own Milky Way galaxy. To their great surprise, they found a substantial signal, coming from all directions in the sky, unwelcome 'noise' so large it actually overwhelmed the signal they were hoping to study. They eventually realized what it was they had found.

---

* You might be unfamiliar with the notion that radiation can have a 'temperature'. In fact, this refers to the temperature of the object that is emitting the radiation, typically spanning a range of frequencies. In this case, the 'object' is the collection of atomic nuclei and electrons, which would have had a temperature around 3000 kelvin at the moment of recombination. Subsequent expansion of the universe then cooled this radiation to the temperature it has today. In the Celsius (or centigrade) temperature scale water freezes at 0°C and boils at 100°C. In the Kelvin temperature scale zero kelvin corresponds to absolute zero (the ultimate zero on the temperature scale), or about −273.15°C.

The cosmic background consists of microwaves and infrared radiation with an average temperature around 2.7 kelvin, almost three degrees above absolute zero. Not that much was known about it in the late 1970s, but cosmology was by now a 'proper' science.

Lee Smolin never took a formal class in physical cosmology, but instead picked up what he needed from independent study. On graduating from Hampshire College he moved to Harvard to study for a Ph.D., and took courses on quantum field theory and the standard model, group theory, and a 'special topics' course on quantum gravity taught by Stanley Deser.* He also took reading courses on advanced general relativity with Raoul Bott and Steven Weinberg, based on the monograph *The Large Scale Structure of Space-time*, published in 1973 by Stephen Hawking and George Ellis. In 1977 Weinberg published a popular book called *The First Three Minutes*, which quickly became a bestseller. The Big Bang was firmly on the map.

But it was also pretty clear that, although the notion of the Big Bang origin of the universe was now broadly accepted, the standard Big Bang theory was beset with difficulties.

It's worthwhile pausing here to reflect on what we might expect such a theory to predict, or at least list the things with which the theory should be consistent. We know the universe is big, and we know it is expanding at a certain rate. As far as we can tell this is a universe whose spacetime is locally flat, which is why we're obliged to learn Euclidean geometry in school. This doesn't prevent the universe as a whole from possessing a spherical geometry, just as we know that from our perspective watching

---

* Smolin's first research project (and his first published paper) was concerned with building a discrete theory of quantum gravity based on techniques developed to study quantum chromodynamics. We'll return to this approach in Chapter 8.

the game at the weekend, the playing field is decidedly level, even though the Earth is round.

We also prefer a theory which can in some way accommodate the things that we see in the universe: stars and galaxies and planets and, obviously, us. There's more. As we stand, gazing in awe at the night sky, we might conclude that the universe appears much the same in different directions. We see stars, smudges of light that we now know are distant galaxies, and empty space. We marvel at this uniformity. But, as the small child clutching your hand will quickly tell you, the pattern of points of light over here looks very different compared with the pattern of points of light over there. We must own up to the fact that the universe has a *structure*, and we'd like to see this reflected in the theory, too.

In 1917, Einstein had argued for a universe that is finite but unbounded. Five years later Russian physicist and mathematician Alexander Friedmann presented three different kinds of possible solutions of Einstein's field equations (without the cosmological constant), and which describe different kinds of universe.

An obvious consequence of a finite universe is that its post Big Bang evolution (and ultimate fate) is then determined by the amount of mass-energy in it. A very high density of mass-energy slows the rate of expansion, applying a kind of cosmic brake, eventually causing the universe to collapse back in on itself. Such a universe is said to be 'closed'. The local spacetime in a closed universe is positively curved, like the surface of a sphere, on which the angles of a triangle add up to more than 180°.

In contrast, an 'open' universe contains insufficient mass-energy to prevent it from expanding forever, leading eventually to a 'heat death' as all the matter and energy in it becomes spread uniformly thinly. Such a universe would have a local spacetime which is negatively curved, like the surface of a saddle, on which the angles of a triangle add up to less than 180°.

Our universe would appear to be consistent with Friedmann's third example, neither closed nor open, in which local spacetime is perfectly flat. But such a universe must then contain a critical or 'Goldilocks' density of mass-energy—neither too much nor too little—to balance precisely the expansion rate. This might be okay in principle but it implies an incredible balancing act, and an astonishing level of fine-tuning. Disturb this delicate balance by just one part in a hundred thousand billion and either the universe would have collapsed long ago or it would have flown apart so fast that galaxies would have never formed.

This is called the 'flatness problem'.*

Measurements of the cosmic background radiation that were available in the 1970s confirmed its astonishing uniformity in temperature across the sky. But such uniformity was hard to explain. Place a hot object in contact with a cool object, and we know that heat will flow from one to the other and the temperatures of both will equalize. But the very last moment that radiation was exchanged with matter was at the moment of recombination, estimated to have occurred about 380,000 years after the Big Bang. There was really no good reason to suppose that the matter and radiation present in the universe at this moment was distributed uniformly. And if it wasn't, there was every reason to expect that the temperature of the cosmic background radiation should be different when we look in different directions across the sky. But it isn't.

This is called the 'horizon problem'.

Young American postdoctoral researcher Alan Guth and his colleague Henry Tye at Cornell University in Ithaca, New York, attempted to resolve some of these (and other) puzzles in 1979. As Guth explored the wider effects of one possible solution, he

---

* Although some theorists insist that it's not a problem at all.

realized that pumping a very large amount of energy into space-time itself at the very earliest moments after the birth of the universe results in a short burst of extraordinary exponential expansion. 'I do not remember ever trying to invent a name for this extraordinary phenomenon of exponential expansion,' he later wrote, 'but my diary shows that by the end of December [1979] I had begun to call it *inflation*.'[7]

In inflationary cosmological models, it really makes no difference how much mass-energy the early universe contained, or whether this was equal to the critical density required to produce a flat spacetime. It doesn't matter what the initial rate of expansion was. It doesn't even matter whether inflation is applied to the whole of the early universe or just one small bubble of spacetime within this. Proponents of inflation theory argue that no matter what the shape of spacetime prior to inflation, when inflation was done *flat* spacetime was inevitable.

Not everyone agrees, and cosmic inflation is today the subject of some lively debate. The truth is that inflation really doesn't solve the flatness problem; it simply pushes the problem back to the initial conditions that prevailed at the moment of the Big Bang. The fine-tuning might no longer be all about the density required for a flat spacetime, but it is still implied in the selection of initial conditions. We'll be returning to this point in Chapter 14.

Inflation also suggested a solution for the horizon problem. The argument goes that at the onset of inflation, the universe was so small that every part of it was indeed in contact with every other part. Uniformity prevailed, and this uniformity was carried through to the larger universe when inflation was done. At the moment of recombination all the matter and radiation in the universe was uniformly distributed (with one caveat, which I'll explain soon).

A standard model of Big Bang cosmology could now be constructed using a spacetime metric which provides an exact solution to Einstein's gravitational field equations, and which describes a uniform, expanding universe. It is known as the Friedmann–Lemaître–Robertson–Walker (FLRW) metric, named for Friedmann, Belgian theorist (and ordained priest) Georges Lemaître, American physicist Howard Robertson, and English theorist Arthur Walker.

To this model we add so-called 'slow roll' inflation. This works rather differently compared with Guth's original mechanism. It is triggered by a specific kind of quantum field called the *inflaton* field (which acts in much the same way as the Higgs field).* This arguably provides potential solutions for the flatness and horizon problems, with the caveat that not all physicists agree that inflation is really necessary, or desirable.

But the theorists then hit a major roadblock. Cosmic inflation could explain why local spacetime is flat, irrespective of the amount of mass-energy in the universe at its onset. However, there was no getting away from the fact that the amount of mass-energy still must balance the current rate of expansion, which could be estimated from Hubble's Law. The trouble was that when astronomers looked at the amount of mass-energy in the visible universe, there just wasn't enough.

And there was no way in this model to understand how stars and galaxies could have formed.

But there were some clues. The shapes of spiral galaxies like Andromeda betray the fact that they are rotating, trailing swirls of stars, dust, and gas from their spiral arms. The population of stars in such a galaxy is most dense at the centre, suggesting that the gravitational field (the curvature of spacetime) is strongest here. Stars close to the centre are expected to rotate much faster than stars lying further out, where the gravitational field is weaker.

---

* Some theorists have recently advanced arguments that the inflaton field and the Higgs field are one and the same.

But when astronomers measured the rotation speeds at the edge of a spiral galaxy they found them to be much higher than predicted based on the mass of the galaxy's visible stars.

The simplest explanation is that there is much more matter in the galaxy than its visible stars would suggest. Each galaxy is presumed to sit at the centre of a 'halo' of missing or invisible matter, which now goes by the name 'dark matter'. This cannot be ordinary matter that is for some reason simply not visible, as any kind of ordinary matter should interact with electromagnetic radiation and so be 'visible' in some part of the spectrum. This is matter unlike anything to be found in the standard model of particle physics, making itself known only through its influence on the gravitational field. We have absolutely no idea what it is.

Dark matter is utterly mysterious, but nevertheless profoundly important. According to a mechanism first devised by Cambridge astrophysicists Simon White and Martin Rees in 1978, dark matter halos are drawn together by their gravity, concentrating ordinary visible matter at their centres.[8] Eventually, a threshold density of visible matter is reached. The first stars are ignited and galaxies form, a few hundred million years after the Big Bang.

But a universe that starts out with a perfectly uniform distribution of dark matter, lightly sprinkled with ordinary visible matter, goes nowhere very fast. In such a universe the gravitational field would also be perfectly uniform; all the matter would be pulled equally in all directions and so frozen in place. There would be nothing to drive the formation of ever-larger dark matter halos. Such a universe could never give rise to the large-scale structure of galaxies, formed along 'strings' and 'walls', surrounding great 'voids', that is so characteristic of our own.

It wouldn't take much of a disturbance to set the ball rolling. Cosmologists estimated that a small inhomogeneity in the distribution of dark matter in the early universe of as little as one part in one hundred thousand would be sufficient.

How could such small inhomogeneities have arisen? Not long after Guth had published his ideas on cosmic inflation, several authors—including Guth himself, Stephen Hawking, Russian theorist Alexei Starobinsky, and others—suggested that the source of the small inhomogeneities could be traced to *quantum fluctuations* of the inflaton field, amplified to cosmic scales by inflation.

If this is correct it is quite an extraordinary conclusion. We owe the large-scale structure of the visible universe to random quantum fluctuations that occurred at the very earliest moments after the Big Bang.

Although very small, such inhomogeneities would have left some telltale signs in the cosmic background radiation, like a bloody thumbprint left at a cosmic crime scene. A succession of satellite surveys—COBE (launched in 1989), WMAP (2001), and Planck (2009)—has now mapped this radiation in ever more exquisite detail (Figure 14).* These maps reveal subtle temperature variations, of as little as a few hundred *millionths* of a degree or—guess what?—one part in one hundred thousand. The hotspots show where, at the moment of recombination, there was a slightly higher density of matter, which would become the seeds for the accumulation of dark matter and eventually star and galaxy formation. Cold spots show where there was a lower density of matter, which would become voids.

* COBE stands for COsmic Background Explorer. WMAP stands for Wilkinson Microwave Anisotropy Probe, named for David Wilkinson, a member of the COBE team and leader of the design team for WMAP, who died in 2002 after a long battle with cancer. The Planck satellite is named for German physicist Max Planck.

Figure 14. All-sky maps of the small temperature variations in the cosmic background radiation derived from data obtained from the COBE, WMAP (9-year results), and Planck satellites. The temperature variations are on the order of ±200 millionths of a degree and are shown as false-colour differences. The angular resolution of this map has increased dramatically with successive missions.

Analysis of these maps provides much (though certainly not all) of the observational evidence on which theories of the origin and evolution of the universe are constructed.

The discovery of dark matter (whatever it is) didn't resolve the accounting problem. Even though it is understood that there appears to be more than five times as much dark matter as visible matter in the universe, this is still not enough to balance the expansion rate. The universe is still desperately short of mass-energy, by as much as seventy per cent of the total needed. Theorists began to mutter to themselves in dark corners. They had some suspicions where this was heading.

In a largely forgotten paper published in an obscure journal in 1933, Lemaître had suggested that the universe is expanding because empty spacetime is not, in fact, empty.[9]

Einstein had introduced the cosmological term on the left-hand side of his gravitational field equations, as a modification of spacetime. But move the term across to the right-hand side of the equation and it now represents a *positive* contribution to the total mass-energy of the universe. This is not mass-energy associated with ordinary matter or dark matter. It implies that empty spacetime has an energy, sometimes called *vacuum energy*. In fact, Λ is directly proportional to the *density* of the vacuum energy, the amount of energy per unit volume of 'empty' spacetime.

In a grand failure of imagination, the vacuum energy has become more familiarly known as 'dark energy'.

A universe containing dark energy would be expected to have a different expansion history compared with a universe containing only matter. This might not seem to help much, because surely we can't look back in time to determine what the expansion history really is.

Well, actually, we can. The speed of light is finite and fixed, so when we look at events in very distant parts of the universe we see these unfold as they would have appeared at the time the light was emitted. Light from the Sun shows us how it appeared about eight minutes ago, which is how long it takes for the light to travel 150 million kilometres to the Earth. Light from the Andromeda Galaxy shows us how it was 2.5 million years ago. Looking at distant events allows us to look back into the past.

But the most distant galaxies are also the faintest, making it very difficult to measure their distances accurately. That is, until one of its stars explodes spectacularly in a supernova. Then, for a brief time, the whole galaxy is lit up.

In 1998, two independent groups of astronomers reported the results of measurements on very distant galaxies illuminated by a certain type of supernova. The results reported by both groups suggest that, contrary to the expectations that prevailed at the time, we live in a universe in which the expansion of spacetime is actually *accelerating*. Dark energy is the most obvious explanation.

And, as dark energy is identical to a cosmological constant, Einstein's infamous factor is back in the equations of the universe. This accounts for the 'L' (for Lemaître) in the FLRW metric. However, it's unlikely that Einstein would have been pleased by this turn of events. After all, he introduced the cosmological term in an attempt to keep the universe still. And yet here it is accelerating the rate of expansion.

Some consensus has now gathered around a version of Big Bang cosmology called variously the 'concordance' model, the 'standard model of Big Bang cosmology', or the Λ-CDM model, where Λ stands for the cosmological constant and CDM stands for 'cold, dark matter'.*

---

* Don't read too much into my use of the word 'consensus'—several aspects of the model are hotly disputed. This is hardly surprising for a model constituted largely by things we know next-to-nothing about.

The Λ-CDM model has six parameters which are adjusted to ensure consistency with observations (such as, for example, the temperature variations in the cosmic background radiation, and the supernova measurements, among others). The best-fit results from analysis of the most recent Planck satellite data published in February 2015 suggest that the universe began 13.8 billion years ago. Dark energy accounts for about 69.1 per cent of the mass-energy density of the universe and dark matter 26.0 per cent. Ordinary visible matter, or what we used to think of as 'the universe' not so very long ago, accounts for just 4.9 per cent.

For all our ignorance of the 95.1 per cent, what we do know is still utterly extraordinary. 'We can reconstruct in detail the history of the universe, starting with its initial hot, compressed state,' Rovelli exulted. 'We know how atoms, elements, galaxies and stars formed, and how the universe as we see it today developed.'

To go beyond what we know in the hope of shedding some light on what we don't, we first need to take a deep breath. We must now try to work out how we might go about finishing the revolutionary transformations in our understanding of space and time, matter and energy that were begun in the previous century by Einstein. 'There is a point', Rovelli continues, 'at which [the densities of matter and energy] reach the Planck scale: 14 billion years ago. At that point, the equations of general relativity are no longer valid, because it is no longer possible to ignore quantum mechanics. We enter into the realm of quantum gravity.'[10]

PART II

FORMULATION

6

TO GET THERE I WOULDN'T START FROM HERE

The standard models of particle physics and Big Bang cosmology are extraordinary triumphs of the human intellect, of which we should be justly proud. The first describes the properties and behaviour of matter and radiation in quite incredible detail—since the discovery of the Higgs boson there is *no* observation or experimental result in high-energy particle physics that it can't explain (or, at least, accommodate). The second describes the large-scale structure and evolution of the universe. It provides answers to some of the biggest of our 'big questions': answers that, not so very long ago, we might have felt compelled to seek out in sacred texts.

But whilst there's ample cause for celebration, we shouldn't get too carried away. These theories are also riddled with explanatory holes. There's an awful lot they can't yet tell us about how the physical world is put together. Scientists aspire to theories that ideally assume little or (even better) *no* knowledge derived from experience. The standard models of particle physics and Big Bang cosmology fall far short of this ideal. Despite their enormous successes, there is a sense in which they are not very *satisfying*.

Even if we're prepared to draw a veil over some of the more discomforting things that quantum theory says about the nature of physical reality, we must come to terms with what the quantum field theories of the standard model of particle physics *don't* say. These theories can't tell us anything at all about the strengths of the interactions between the elementary matter and force particles and the Higgs field, so they cannot be used to calculate the masses of these particles from 'first principles'. Instead we must measure the masses experimentally and insert them into the equations by hand. Similarly, we have no way of knowing the relative strengths of the forces that act between the matter particles, except to measure them.

All the particles in the standard model have antimatter counterparts. Antiparticles possess the same masses as their matter equivalents but have opposite electrical charges, such as the electron (e^-) and antielectron (or positron, e^+). When they collide, particles and antiparticles annihilate to produce high-energy photons. But, provided they are kept separate from matter, the antiparticles would appear to be perfectly stable.

But the visible universe appears to be made entirely of matter, not antimatter or a mixture. If equal amounts of matter and antimatter were produced in the first few moments following the Big Bang, as seems reasonable, why did it not all annihilate, leaving a universe full of light but devoid of matter? It would seem that, by chance or necessity, as the early universe evolved the balance was tipped ever so slightly in favour of matter particles. Currently accepted theories give no clue as to why this might have happened.

Then there's the puzzle of dark matter, an essential ingredient in the standard model of Big Bang cosmology, but which is entirely absent from the standard model of particle physics. We have lots of speculative ideas, and several experimental searches are underway, but at the present time it would seem we have absolutely no way of knowing what it is. Given that we know of it only through the effects it has on matter and radiation that we can detect, we can't even be sure that dark matter really exists.

The situation with dark energy, the energy of 'empty' space-time, is certainly no better and is in some ways much worse. The current value of the cosmological constant derived from the standard Big Bang model implies a density of vacuum energy on the order of a millionth of a billionth (10^{-15}) of a joule per cubic centimetre. Now, you may not be entirely familiar with the joule as a unit of energy (it is related to the more familiar caloric), but all you really need to do here is focus on the number.

We might be tempted to think that this energy density derives from quantum fluctuations in empty space.* The calculation is tricky, but theorists have made some headway by finding ways to 'regularize' it, applying an arbitrary cut-off, deleting the highest-energy contributions. Taken at face value, this suggests a quantum field contribution to the vacuum energy density on the order of 10^{105} joules per cubic centimetre, a result that differs from what we observe by a staggering hundred billion billion googol (10^{120}). We no longer argue about whether we need a cosmological constant; now, we argue about why it should be so *small*.

Finally, we must confront the elephant in the room. The standard model of particle physics is a collection of quantum field theories which, though they meet the demands of special relativity, still assume a background spacetime. They *pre-suppose* a fabric.

---

* Strictly speaking, these are not fluctuations in time but refer to a related phenomenon—the lowest energy (vacuum) states of quantum fields cannot have zero energy (for reasons already discussed in Chapter 4) but must instead possess an average which is called the zero-point energy. The sum of the zero-point energies of all the different quantum fields then provides a quantum theoretical estimate of the density of vacuum energy.

They are formulated in a way that describes the properties and behaviour of microscopic, quantum-sized objects *in* an absolute spacetime 'container' arguably not so very different in nature from the absolute space and time of Newtonian physics. Such a spacetime is *passive*, a silent witness to the unfolding quantum physics.

In contrast, the standard model of Big Bang cosmology is constructed from Einstein's general theory of relativity, which is background-independent. It does not pre-suppose a fabric. Instead, spacetime emerges as a dynamical variable of the theory. Here spacetime is *active*; it is acted on by mass-energy and acts on mass-energy, and so actively *participates* in the physics.

For sure, general relativity is a theory of the gravitational field (spacetime), but this is a classical, not a quantum, field. Just like Maxwell's classical electromagnetic field, it simply can't exhibit the characteristic fluctuations, disturbances, or excitations which we would then interpret as the quanta of the field. There are no 'field particles' as such in general relativity. In the case of the gravitational field such quanta are at present entirely hypothetical. They are called *gravitons*, a name coined by Dirac in 1959.

Einstein was all-too-aware of the contradiction. As we saw in Chapter 2, he presented his general theory of relativity in a series of lectures delivered to the Prussian Academy of Sciences in Berlin, culminating in a final, triumphant lecture on 25 November 1915. Yet within a few short months he was back, explaining to the academy that his new theory of gravitation might need to be modified:[1]

> Due to electron motion inside the atom, the latter should radiate gravitational, as well as electromagnetic energy, if only a negligible amount. Since nothing like this should happen in nature, the quantum theory should, it seems, modify not only Maxwell's electrodynamics but also the new theory of gravitation.

Bringing quantum theory and general relativity together gives us a quantum field theory of the gravitational force, or a *quantum theory of gravity*. It is a quantum theory of spacetime itself.

You might wonder why, with such an obviously missing ingredient, the current standard models of particle physics and Big Bang cosmology continue to work as well as they do. The simple answer is that the forces involved in both models work on enormously different scales. Quantum theory applies at microscopic subnuclear, subatomic, atomic, and molecular scales, and the forces involved are relatively strong. In contrast, general relativity applies most noticeably at the scale of macroscopic objects, such as people and rockets, planets, stars, galaxies, and the entire universe. Compared with the forces involved in particle physics, the 'force' of gravity caused by the curvature of spacetime is extremely weak.

Your personal experience of gravity might lead you to doubt this, especially if you've ever fallen down and grazed a knee. So here's a simple experiment to demonstrate the relative strengths of gravity and electromagnetism that you can do at home. Just follow these simple instructions:

1. Place a metal paper clip on the table in front of you.
2. Gradually lower a handy-sized magnet above it.
3. At some point, the paper clip will be pulled upwards from the table by the magnetic force and will stick to the bottom of the magnet.

Congratulations. You've just demonstrated to yourself that in the tussle for the paper clip, the force generated by the small magnet in your hand is stronger than the *gravitational pull of the entire Earth*.

This great difference in the strengths of nature's forces is both a blessing and a curse. It's a blessing because it means that we can safely neglect the curvature of spacetime and the resulting gravitational effects when considering interactions involving elementary particles. The masses involved are too small and gravity too weak to make any real difference (and indeed, Einstein said: 'if only a negligible amount'). By the same token, we can use general relativity to describe the motions of planets, stars, galaxies, and the large-scale structure of the universe without needing to take into account the forces acting on their constituent atoms, quarks, and electrons.

It's a curse because the distance scale at which any quantum effects of gravity might be expected to become important is painfully small, the associated energy scale painfully large, substantially well beyond the reach of any experiment we might devise, now or in the future.* This is the *Planck scale*, which Carlo Rovelli mentioned at the end of the previous chapter. This scale is defined by combinations of a small selection of fundamental physical constants.[2] The Planck length is on the order of 10^{-35} metres, or a billion billionth of the radius of a proton. The Planck energy is a million billion times larger than the energies of the proton–proton collisions created in the Large Hadron Collider at CERN, so far the highest-energy particle collisions ever created on Earth.

We presume that the universe would have experienced the Planck scale in the very earliest moments of the Big Bang, but that happened 13.8 billion years ago. Experimental physics is never going to be able to get even remotely close to this scale, so we might ask ourselves: Why worry? We apply quantum theory to elementary particles and we apply general relativity to big stuff and the universe, and everything works fine. Well, except

---

* In science there are always caveats. See Chapter 14 for some further discussion.

that we know it doesn't, really, and we understand that there ought to be some overarching theory that should, in principle, describe everything. Also, we have deep suspicions that the very different ways in which we treat the fabric of spacetime in general relativity and quantum mechanics might be at least partly to blame.

Feynman, for one, didn't believe it would be necessary to go to such ridiculous energies, preferring to think instead that exquisitely precise measurements might one day make quantum gravitational effects accessible to inspection in the laboratory.

Either way, a quantum theory of gravity is so obviously missing from our description of nature that, to many theorists, the temptation to construct one is simply irresistible. Attempts were begun in 1930, independently by Belgian physicist Leon Rosenfeld, Soviet physicist Matvei Bronstein, and Pauli. These were rather premature and quickly frustrated by the kinds of problems that would plague the early development of quantum field theory.

It was realized that there are two principal roads that can be taken, two ways of getting there (a quantum theory of gravity) from here (quantum theory and general relativity).

Now, to a certain extent, physical mechanics is much like financial accounting. Whereas accountants seek to balance books that detail the flows of money among a business, its customers, and its suppliers, physicists seek equations that balance the flow of energy from one physical system or situation to another. There is a fundamental law of the universe that energy is conserved and, although it can be manifested through many different particles and forces, it comes in two basic types: kinetic and potential.

Kinetic energy is simply the energy associated with motion. Potential energy is slightly more obscure. It represents the energy that is stored or latent in a physical system. A good way of thinking about potential energy is to imagine a pendulum, swinging back and forth. It swings to the left, sweeps out an arc, and slows down as it rises to its maximum displacement. When the pendulum halts for the briefest of instants as its motion changes direction, all its energy is in the form of potential energy. Pulled by gravity, it then falls back and swings to the right, picking up speed as the potential energy is converted back into the kinetic energy of motion. There's nothing in principle to prevent this going on for-ever, but energy slowly 'leaks' away through mechanical friction and air resistance, and eventually the pendulum comes to a halt.

It might help to know that in elaborations of classical mechan-ics in the years after Newton, physicists established a direct relationship between potential energy and Newton's concept of force. At the top of its swing, the potential energy *curve*— the variation of the potential energy of the pendulum with displacement—is at its steepest. And this is where the force acting on the pendulum is strongest.

If kinetic energy is the income statement, then potential energy is the balance sheet. To account properly for the health of a business, we need to include both.* In a mechanical system the *sum* of the kinetic and potential energies is called the system *Hamiltonian*, named for nineteenth-century Irish physicist and mathematician William Rowan Hamilton.

Dirac devised a successful recipe for developing quantum field theories. Start with a classical field (such as Maxwell's electro-magnetic field) and deduce the equation of its total energy—the Hamiltonian. Then season it by adding some quantum flavour, using a couple of mathematical tricks that recognize how clas-sical properties translate to their quantum equivalents, and

---

* Readers who know something of business accounting will also know that a business lives or dies on the basis of its cash flow, and both income statement and balance sheet are required to determine this.

which introduce field quanta which can be created and destroyed. All this is done whilst preserving the formal structure of the classical theory.

This recipe is called *canonical quantization*. The result is a good first approximation to a quantum field theory equivalent to the classical theory. This approach is, in fact, the subject of Dirac's *Lectures on Quantum Mechanics*.

The first road to a quantum theory of gravity makes use of Dirac's recipe. We start with general relativity and cast Einstein's gravitational field equations into a set of classical Hamiltonian equations, as a kind of stepping stone. This dynamical representation of general relativity is sometimes referred to as *geometrodynamics*, by analogy with electrodynamics. We then season using canonical quantization, as required, to give a quantum geometrodynamics, analogous to QED. This is called the *canonical approach to quantum gravity*.

But the active nature of spacetime in general relativity makes this a very different prospect when compared with the passive spacetime of electromagnetism. In general relativity there is no meaningful 'here' and 'there' or 'now' and 'then'. The theory rather deals with spacetime *intervals*. We might be tempted to imagine that, in a dynamical version of general relativity—which places such emphasis on treating space and time on an equal footing—a correct representation ought then to involve changes in all four spacetime intervals, three of space and one of time.

Instead, when Dirac himself followed this road in the late 1950s, he discovered that in a so-called constrained Hamiltonian reformulation of general relativity, the dynamics are governed by only three of the four spacetime intervals. These hold all the information about the geometrical relationships that dictate what will happen to the mass-energy and, to all intents and purposes, look like intervals in three spatial dimensions.

Reformulating general relativity in this way had unpicked four-dimensional spacetime to yield a '3 + 1' structure. Time had not exactly disappeared, but it had become rather mysterious and elusive. In fact, the changing relationships in spatial geometry can be interpreted in this formulation as different instants in time. Applying the techniques of canonical quantization to this kind of formulation means that it is *space* that is quantized, not spacetime. 'This result', Dirac declared, 'has led me to doubt how fundamental the four-dimensional requirement in physics is.'[3]

Although Dirac understood the mathematics well enough, he made little progress towards a solution, and in his *Lectures on Quantum Mechanics* he summarized the situation like this:[4]

> People have in recent years worked to some extent on bringing the gravitational field into quantum theory, but I think that the main object of this work was the hope that bringing in the gravitational field might help to solve some of the difficulties. As far as one can see at present, that hope is not realized, and bringing in the gravitational field seems to add to the difficulties rather than remove them.

In 1959, Charles Misner, working together with fellow Americans Richard Arnowitt and Stanley Deser, published an elaboration of the constrained Hamiltonian formulation of general relativity, which became commonly known as the ADM formulation. Working at the Institute for Advanced Study in Princeton, New Jersey, in 1966 American theorist Bryce DeWitt used the ADM formulation to develop a canonical quantum field theory for a simple Friedmann universe.

This was the first example of a quantum theory of cosmology. The notion of a 'wavefunction of the universe' entered the scientific literature for the first time. DeWitt found that such a wavefunction depends only on spatial geometry: time still remained stubbornly elusive. The wavefunction describes a stationary, model universe with zero total energy. He wrote: 'one therefore comes to the conclusion that nothing ever happens in quantum [geometrodynamics], that the quantum theory can never yield anything but a static picture of the world.'[5] To get anything to happen in this universe he needed to find a way to reintroduce time, by adding together many different wavefunctions representing different quantum states of space in a superposition, and then letting this evolve.

DeWitt had no clear sense of how the wavefunction of the universe should be interpreted and, in mitigation, cited historical precedent. Schrödinger had faced a similar challenge when confronted with the notion of the wavefunction of an electron in a hydrogen atom. In that case, the wavefunction, and specifically Born's interpretation in terms of 'waves of probability', had been subsumed into the Copenhagen interpretation of quantum theory. But, as DeWitt now argued, in a quantum theory of gravity the Copenhagen interpretation could offer no help whatsoever:[6]

> The Copenhagen view depends on the assumed *a priori* existence of a classical level to which all questions of observation may ultimately be referred. Here, however, the whole universe is the object of inspection; there is no classical vantage point, and hence the interpretation question must be re-argued from the beginning.

If we assume that everything there is is *inside* the universe, then there can be no classical 'measuring device' sitting outside whose purpose is to collapse the wavefunction of the universe. This led DeWitt to campaign vigorously for the alternative many worlds interpretation of quantum theory, which eliminates the need to invoke the collapse of the wavefunction but leaves us pondering the fates of all those Gwyneth Paltrows.

DeWitt had derived his equation by combining Einstein's general relativity with Schrödinger's version of wave mechanics, so he called it the Einstein–Schrödinger equation.* Wheeler had done much to publicize DeWitt's result, and so everybody else called it the Wheeler–DeWitt equation.

This was not the end, however. At best, the Wheeler–DeWitt equation represents a useful staging post on the journey to a quantum theory of gravity, one that is positioned rather far from the ultimate destination. It actually represents an infinite set of equations, and is beset with technical difficulties that prevented theorists from deducing any solutions. Talking generally about the wavefunction of the universe was all very well, but it was obvious that nobody was going to be in a position to write down the form for such a wavefunction anytime soon. Rovelli later admitted: '[The Wheeler–DeWitt equation] presented many difficulties: the mathematics on which it was based was not well defined and its physical significance was very obscure.'[7]

Attention waned. The late 1960s and 1970s was in any case a golden era for high-energy particle physics. By the end of this period the standard model of particle physics was more or less in place, waiting for collider experiments to hunt down the remaining missing particles and fill the gaps.

And the particle theorists had come to see things quite differently.

The second road to a quantum theory of gravity approaches from the direction of quantum field theory and seeks to find ways to make it conform to Einstein's principle of general covariance. In other words, start with quantum field theory and try to make

---

* In private he would often refer to it as 'that damned equation'.

it background-independent. This is called the *covariant approach to quantum gravity*.

In this approach primacy is given to the quantum field theory techniques that were proving to be so successful in building the standard model of particle physics. It involves splitting the spacetime metric of general relativity into two components. The first is a passive, typically flat or so-called Minkowski spacetime, without a gravitational field. The second component is then taken to describe tiny deviations from flatness which represent fluctuations of the graviton field. This field is then quantized, and the gravitons that result propagate on the flat spacetime. Early results seemed promising. The gravitons thus produced were found to possess the properties expected of them.

Feynman himself weighed in, adapting the techniques that had been used so successfully in QED and applying them to gravity. DeWitt also made important contributions, as theorists tried to compute all the most important radiative corrections necessary to renormalize the theory. But it was quickly discovered that a quantum theory of gravity built this way just couldn't be renormalized. Several attempts were made to fix this but, by the late 1970s, none had succeeded.

Feynman was not greatly downhearted, however:[8]

> I am not dissatisfied with my attempt to put gravity and quantum mechanics together. I accept whatever consequences that this... produces, mainly that it can't be renormalized... I always thought that it just meant that we've gone too far: that when we go to very short distances the world is very different.

But with the priority firmly fixed on the quantum field, the covariant approach sacrifices the deep connection between gravity and spacetime geometry that lies at the very heart of general relativity. And those physicists who had devoted their careers to research on general relativity did not like the methods of renormalization that the quantum field theorists had been obliged to resort to.

Although some theorists acknowledged the benefits of pursuing multiple lines of research, the conceptual divisions and lack of common ground between the research groups pursuing these two roads proved to be too great. The emphasis on geometry in the canonical approach 'drove a wedge between general relativity and the theory of elementary particles'.[9] But the particle theorists' extraordinary success and a burgeoning public interest in this kind of science led them to develop the conviction that the covariant approach was the only sensible road to take.

This was the situation that confronted Lee Smolin and Carlo Rovelli as they entered their final years of undergraduate study.

Reading Einstein's *Autobiographical Notes* had persuaded the young Smolin not only to pursue a career as a theoretical physicist but also to devote himself to the development of a quantum theory of gravity. For his doctorate at Harvard he wanted to study quantum field theory with the intention to steal the techniques that had been used so successfully to build the standard model of particle physics and apply them to quantum gravity. But his teachers advised him that 'only fools work on this problem'.[10] Undaunted, he argued the case with his thesis advisor, Sydney Coleman, who tried to persuade him to choose a different topic. Smolin remained stubborn. There wasn't much of an argument. Coleman said he would give Smolin two years to come up with something or he would suggest an alternative problem in QCD.

But he also did Smolin another great favour. He asked Stanley Deser to share responsibility for supervising Smolin's research efforts.

Rovelli learned about the problems of quantum gravity in his final year of undergraduate study at the University of Bologna.

He was handed a copy of a paper, written by English physicist Chris Isham at Imperial College, London, which had been published in the proceedings of a symposium on quantum gravity that had been held in February 1974 at Britain's Atomic Energy Research Establishment at Harwell, near Oxford.

It was clear from Isham's introductory paper that the quantum theory of gravity was beset with seemingly intractable problems and the situation was hopelessly confused. But, filled with the rash courage of youth, the philosopher in Rovelli was nevertheless strongly attracted to the idea of studying the concepts of space and time, perched right on the very edge of scientific understanding.

Rovelli's teachers tried to discourage him, arguing that spending years of study on such an esoteric and unproductive subject would mean that he would surely never find a job at the end of it. But, like Smolin, he was unperturbed: 'often the only result of the cautious advice of adults is that it strengthens the cheerful stubbornness of the young'.[11]

Together with some student friends, he edited a book about the events which had unfolded in Bologna in March 1977, in which a young militant had been killed and Radio Alice had been forced off the air, followed by a few rather feverish days of demonstrations, barricades, and rioting. Such 'crimes of opinion' led to Rovelli's arrest and earned him a beating at the local police station.

Although preoccupied with student politics, he continued to think about the problems of quantum gravity and shortly afterwards moved to the University of Padua to study for a doctorate. He was fortunate to find a thesis adviser who was more than happy to leave him alone, allowing him the freedom to study whatever he wanted. As his fellow graduate students began to publish their first research papers, he kept his head down.

At this stage he just wanted to understand.

7

A GIFT FROM THE DEVIL'S GRANDMOTHER

In truth, a theoretical physicist never really starts with a blank sheet of paper, or goes back to a drawing board that is completely empty. The landscape of possible mathematical structures that might be used to describe the physics of elementary particles or the universe is broad but it's not infinite. And although fortune tends to favour the rebellious, there is still a strong incentive to stick with structures that we already know. Revolutions are always fought with contemporary weapons.

I don't think it's all that difficult to imagine the way that theory gets done. In the old days, when there was plenty of unexplained empirical data lying around, scientists would try to induce a theory using familiar, mathematically accessible constructions, such as point particles (with all their mass assumed to be concentrated to infinitesimally small points at their centres), extended three-dimensional fields, the motions and vibrations of vortex objects, and so on. They would try to build a theory and see not only if it 'fit' the existing data but also predicted some new and preferably unusual facts that could then be sought in some further experiments or observations.

In situations where new data are a little more difficult to come by, scientists might try introducing some novel concepts and develop a theory which provides a *better explanation* or rationalization of the existing data by plugging one or more of the explanatory gaps. This sometimes involves a complete reinterpretation of existing theoretical concepts: we thought this meant that, only to find that it means something rather different, or (in extreme cases) something else entirely.

Either way, the theorists start with a *big idea*, teased out from attempts to second-guess nature. The theorists' indulge their inner metaphysician, much like the philosophers of ancient Greece. They ask themselves 'what if' questions and reach for already tried-and-trusted constructions to explore potential answers, or even just to see where the mathematical logic will take them.

The incredible fertility of the collection of field theories that make up the standard model of particle physics—and the general excitement that built up around the establishment of this model during the 1970s—suggested to many theorists that this was a really good place to start.

Given that the unification of the weak and electromagnetic forces had been so successfully demonstrated, the next question was really rather obvious. What if we unify the strong colour force, too, in a single 'electronuclear' force? This would imply that *all* the quantum fields and forces that operate inside the atom are really the broken remnants of a much grander, unified field with a higher symmetry.

But let's not stop there. What if there's a symmetry between matter particles (fermions) and force particles (bosons), with every particle in the standard model possessing a 'partner', just as all particles are partnered by their antiparticles? What if the real elementary physical entities are not particles at all, but one, two, or higher-dimensional filaments of energy called *strings*?

Those theorists trying to chase down the secrets of nature are obliged to allow their imaginations to run if not wild, then pretty excitably. Of course, the big ideas must be constrained by the physics they're trying to explain, but nobody drives a revolution in scientific thinking by being timid.

And so this approach inevitably comes with a couple of significant trade-offs.

The theorists are hunting in the dense jungle of complex mathematics. It's easy to get lost (as Einstein did in tensor calculus) and even easier to lose sight of the goal. They set off in hot pursuit of results that must first make *mathematical* sense, bringing in often obscure concepts to which we all later struggle to give physical meaning or interpretation. This doesn't—it cannot—trouble them. They are seeking a structure that *works*, not a structure that is necessarily easy for ordinary mortals to comprehend, no matter how satisfactory that would be (or how easy it would make the later task of popularization).

But the second big trade-off *is* rather troubling and comes down to the way we interpret the word 'theory'. Of course, we use this word very loosely in everyday discourse—I have a theory about why UK citizens voted narrowly to leave the European Union in June 2016 and why Donald Trump was elected as the 45th President of the United States later that year. We can all agree that this is 'just a theory'.

But to scientists, *successful* theories are much more than this. Even though their concepts might sometimes be a bit obscure, these theories still tell us something deeply meaningful about the way nature works. Theories such as Newton's system of mechanics, Darwin's theory of evolution, Einstein's special and general theories of relativity, and quantum mechanics are broadly accepted as contingently 'true' representations of reality and form the foundations on which we seek to understand how the universe has evolved and how we come to find ourselves here, able to theorize about it. Much of what we take for granted in our complex, Western scientific-technical culture depends on the application of a number of reliable scientific theories. We have good, solid reasons to believe in them.

Cell biologist Kenneth R. Miller explained that a scientific theory 'doesn't mean a hunch or a guess. A theory is a system of explanations that ties together a whole bunch of facts. It not only explains those facts, but predicts what you ought to find from other observations and experiments.'[1]

So, we wrap the new big idea around a suitable mathematical structure, but until we can do something useful with it in relation to the facts, it's arguably a scientific *hypothesis*, not a fully fledged theory. 'Theories' that cannot (yet) establish a foundation in empirical data simply do not have the same *status* as theories that can.

Nevertheless, even scientists still use the word 'theory' rather loosely. And as it has become harder and harder over the past few decades to obtain new empirical facts to help anchor excitable speculation more firmly in the real world, as we will soon see some theorists have developed a tendency to reinterpret what they are prepared to regard as 'success', particularly those theorists with rather grandiose ambitions to formulate a so-called *theory of everything*.

Attempts in the early-mid 1970s to build a grand unified theory of the strong colour, weak, and electromagnetic forces illustrate quite nicely how theory development *should* work. Despite appearances, the field theories that make up the standard model of particle physics are quite separate and disconnected. We reach for QCD to study quarks and gluons and the broken field theory of the electroweak force to study weak and electromagnetic interactions between quarks and leptons.

This is all a bit unsatisfactory, and the search began for a field theory that would bring all these fields and forces under one roof. In 1974 American theorists Sheldon Glashow and Howard Georgi thought they had found it. Now, such a grand unified theory is obliged to possess a higher symmetry and more mathematical 'dimensions' than the theories that it devolves into as the symmetry is broken. It was imagined that the grand unified force would first split into the strong colour force and the electroweak force, thought to have occurred about 10^{-35} seconds after the Big Bang, presumably due to some kind of unspecified interaction with a Higgs-like field.* Then the electroweak symmetry would have been broken by interaction with the Higgs field, giving rise to the weak force and electromagnetism, about a trillionth (10^{-12}) of a second after the Big Bang.

At the time this must have seemed like progress, but an inevitable consequence of the higher symmetry is that every particle now has some kind of relationship with every other particle, governed by an 'electronuclear' force carried by hypothetical 'X' bosons. Some of these relationships persist when the electronuclear symmetry is broken, and in the Georgi–Glashow theory this meant that quarks inside the proton become susceptible to a kind of radioactive decay, transforming the proton into a neutral pion and a positron.

'And then I realized that this made the proton, the basic building block of the atom, unstable,' Georgi said. 'At that point I became very depressed and went to bed.'[2]

This is exactly what we want a speculative theory to do: *make a prediction*. Georgi became depressed because he knew that the proton is not unstable. Subsequent experiments, involving the search for telltale proton disintegrations in huge tanks of ultrapure water buried far below ground, showed quite clearly that the proton is more stable—it survives at least ten thousand times longer—than the Georgi–Glashow theory predicts.[†]

* There's a line of argument that suggests this symmetry-breaking was triggered by the inflaton field, leading to the onset of cosmic inflation.

† Italian physicist Carlo Rubbia explained it like this: 'just put half a dozen graduate students a couple of miles underground to watch a large pool of water for five years' (quoted by Peter Woit, *Not Even Wrong*, Vintage, London, 2007, p. 104).

Other approaches were tried, but no real solutions were forth-coming. An annual scientific conference on the subject of grand unified field theories established in 1980 did not survive beyond 1989, as the theorists turned their attentions elsewhere. And this is what we would expect. A theory that makes predictions that can be demonstrated to be false is eventually abandoned. Lessons are learned, and the physics community moves on.

'Perhaps, to make further progress,' Glashow admitted, 'we simply have got to include gravity as well. This was always Einstein's belief, and for the last thirty years of his life it had led him down the Princeton garden path.'[3]

So, we reach for another big idea. Perhaps the theorists were a little too ambitious, trying to build a unified field theory before we've properly understood what it is we're supposed to be unify-ing. After all, the standard model of particle physics is riddled with plenty of problems of its own, and maybe it's best to try to fix these first. One such is called the 'hierarchy problem', particu-larly as it is manifested in the calculation of the mass of the Higgs boson.

The discovery of the Higgs boson in July 2012 was obviously a great triumph. But this was not just about building a particle col-lider big enough to generate the kinds of collision energies needed to bring it out into the open. It was about building a collider that could explore a broad range of energies because, despite its importance, before it was discovered nobody could tell you what the mass of the Higgs ought to be.

This is because the standard quantum-theoretical approach to calculating the mass of the Higgs involves computing radiative corrections to the particle's bare mass, thereby renormalizing it. These corrections involve taking account of all the different pro-cesses that the Higgs boson might undergo as it moves around, including virtual processes involving the production of other particles and their antiparticles for a short time before these recombine. Now, the theory demands that the Higgs boson is obliged to couple to other particles in direct proportion to their masses, so virtual processes involving very heavy particles such as the top quark are expected to make significant contributions to the dressed mass of the Higgs.

To cut a long story short, on this basis the mass of the Higgs is expected to mushroom to the Planck scale. Now, the Planck mass is a little more than 0.02 milligrams, or about ten billion billion times the mass of a proton. That's big enough to see with the naked eye! Clearly, something must be happening to cancel out the contributions from all these radiative corrections, 'tuning' the Higgs mass to the value it was eventually found to possess, about 133 times that of a proton.

There are some fairly obvious potential explanations. As American physicist Stephen Martin explained in 2011: 'The systematic cancellation of the dangerous contributions to [the Higgs mass] can only be bought about by the type of conspiracy that is better known to physicists as a symmetry.'[4]

The symmetry in question is called *supersymmetry*. The first theories of supersymmetry were developed in the early 1970s by a number of Soviet physicists based in Moscow and Kharkov. The theory was independently rediscovered in 1973 by CERN physicists Julius Wess and Bruno Zumino.* Note that this is *not* a grand unified theory. It's probably best to think of supersymmetry as an important stepping stone. If it can be shown that nature is indeed supersymmetric, then some (but not all) of the problems of the present standard model of particle physics can be resolved and the path to a grand unified theory might become a little clearer.

---

* As is typical, the history is a little more complicated than this, with several theorists discovering and then rediscovering the basic principles.

Theories based on the assumption of supersymmetry establish a fundamental spacetime relationship between fermions and bosons. Such theories inevitably proliferate more particles. For example, in the so-called Minimal Supersymmetric Standard Model (MSSM), the simplest application of supersymmetry to the current standard model of particle physics, for every fermion the theory predicts a corresponding supersymmetric fermion (called a sfermion, a shortening of 'scalar'-fermion), which is actually a boson. The partner of the electron is called the selectron. Every quark is partnered by a corresponding squark.

Likewise, for every boson in the current standard model, there is a corresponding supersymmetric boson, called a bosino, which is actually a fermion. Supersymmetric partners of the photon, W and Z particles are the photino, wino, and zino.*

Assuming supersymmetry means that radiative corrections involving massive fermions tend to be cancelled by radiative corrections involving massive sfermions. Now, as a mathematician I have very limited experience and virtually no real ability. But what experience I do have allows me the following insight. If, when grappling with a complex set of mathematical equations, you are able to show that all the unruly terms neatly cancel and the answer is a meaningful number, the result is pure, unalloyed joy. And this is what happens in supersymmetry theories. The end result is that the mass of the Higgs boson can, in principle, be stabilized at a sensible value.

There are precedents for this kind of thing. The existence of matter and antimatter particles reflects just such a symmetry in nature. But the symmetry between matter and antimatter is 'exact'—aside from their different electric charges an electron and its positively charged partner, the positron, behave in much

---

* The wino is pronounced 'weeno', presumably to avoid confusion.

the same way and have precisely the same mass. This cannot be the case for supersymmetry, because if this was an exact symmetry, then we might expect selectrons (to take one example) to have the same mass as electrons, suggesting that selectrons ought to be at least as common as positrons.* If these really did exist, we would surely have found them by now.

The simple fact that no sparticles have ever been observed means that if nature really is supersymmetric, then the supersymmetry must be broken, pushing the mass range of the super-partners beyond reach of every particle collider ever built so far.[†] It's fair to say that we don't have a very good theoretical explanation for how this is supposed to happen.

And the experiments performed at the Large Hadron Collider to date pretty much rule out the simplest supersymmetric theories (such as the MSSM). Now, theories with just one supersymmetry require the introduction of about 120 additional parameters, most of them related to spontaneous supersymmetry-breaking. This is obviously many more than the 20 or so parameters of the standard model we are notionally trying to fix. Theories with more super-symmetries not only suffer a parameter problem, but it becomes virtually impossible to break these supersymmetries spontaneously in ways that will recover the standard model particles that we know.

The assumption of supersymmetry might have helped to eliminate troublesome radiative corrections, but this is no longer seen as the 'natural' solution to the hierarchy problem, and it provides virtually no basis for making any kind of testable prediction.

---

* Another issue is that if the supersymmetry were exact, then all matter particles would be unstable ...

[†] The most recent (March 2017) results from the ATLAS collaboration at CERN's Large Hadron Collider suggest that, if it really does exist, the mass of the gluino, the hypothetical supersymmetric partner of the gluon, must be well above 2,000 times the mass of a proton.

It is somewhat misleading to think that the big idea involved in supersymmetry is all about the pattern of super-partners. Yes, sparticles are an inevitable consequence but the really big idea is the assumption of a fundamental *spacetime symmetry* between fermions and bosons. In fact, supersymmetry is not so much a theory as a *property* of a whole class of theories. There are many different kinds of supersymmetry theories.

And therein lies a clue. If we assume that the spacetime of general relativity is supersymmetric, then we have a theory of *supergravity*. Amongst the earliest examples were versions developed in 1976 by American Daniel Freedman, Dutch physicist Peter van Nieuwenhuizen, and Italian Sergio Ferrara and, independently, by Stanley Deser and Bruno Zumino. These theorists discovered that some of the problems associated with renormalizing a quantum field theory of gravity were somewhat (but not totally) relieved if supersymmetry was assumed. It seemed that the infinite contributions from radiative corrections involving the graviton that had plagued the efforts of Feynman and others could be partly offset by corresponding contributions from the gravitino, the super-partner of the graviton.

Excitement built up around an extended version of supergravity based on eight different kinds of supersymmetry. This theory accommodates not only the graviton, but eight gravitinos and 154 other particles (which, taken at face value, implies that quarks and gluons might not, after all, be elementary particles). For a short time, this was believed to be the 'real deal'. In his 1980 inaugural address as Lucasian Professor of Physics at Cambridge University in England (a position once held by Newton), Stephen Hawking asked if the end was in sight for theoretical physics. At the time the jury was still out on supergravity, but Hawking argued that this was 'the only candidate in sight'.[5]

But, although supersymmetry appeared to promise much, it could not completely eliminate the problems of renormalizability.

The renormalizability of supergravity based on eight supersymmetries remains an open question today, but in the early 1980s the problem appeared insuperable and interest in the theory began to wane.*

Then, in 1982, a young theorist named Amitabha Sen, working as a postdoctoral associate at the Center for Theoretical Physics at the University of Maryland, published a couple of papers that had some in the theory community suddenly sitting up and taking notice.[†]

To help explain what Sen was doing, we need to take a short diversion.

Vectors have a profoundly important role to play in pretty much all of physics. These are physical properties characterized by possession of a certain magnitude (small, medium, large) and direction (pointing this way or that way). Perhaps one of the simplest examples of a vector quantity in physics is linear momentum. In classical mechanics the magnitude of an object's linear momentum is simply given by the object's mass multiplied by its velocity. Its direction is obviously the direction in which the object is moving, from here to there. We typically represent such a vector as an arrow drawn from the object, pointing along the direction of motion, whose length relates to the magnitude: a larger linear momentum vector implies a longer arrow.

Both magnitude *and* direction are key to specifying a vector quantity—just ask any professional tennis player preparing to serve.

In quantum mechanics, vectors take on an even greater, if not elementary, significance. I mentioned electron spin in Chapter 4— the electron can take up two (and only two) different orientations

---

* Radiative corrections involve dealing with 'loops', of varying degrees of complexity. Supergravity proved to be renormalizable for simpler, so-called one- and two-loop corrections, but could not be renormalized for higher-order corrections.

[†] Sen had already published a couple of papers (in 1981 and 1982) as a student at the University of Chicago.

in a magnetic field which we refer to as spin-up and spin-down. These are properties that can be traced to the electron's intrinsic spin angular momentum, and we can treat spin in terms of vectors (each with a magnitude given by $h/4\pi$, where h is Planck's constant) which can 'point' in two different directions.

Although vectors are a little more complicated than scalar (magnitude-only) quantities, we have no real difficulty analysing the mechanics of vectors moving around in a flat, Euclidean space. A vector with a certain magnitude and direction (let's say it's pointing up) might be displaced by some physical force from one location, with coordinates $x_1 y_1 z_1$, to another location $x_2 y_2 z_2$, without changing magnitude or direction, and all's well. But as soon as we switch to a curved space we start to run into problems.

To see why, it's helpful to learn a little about a third-century Chinese invention called the *south-pointing chariot*. This was a two-wheeled chariot bearing a carved wooden figure with an out-stretched arm pointing in one direction. At the beginning of a journey the figure would be manually set to point due south, and an ingenious gear mechanism would then ensure that with every twist and turn of the chariot, the figure would continue to point south (this was long before the invention of the magnetic compass).

Think of the figure as representing a vector. We start at the North Pole, so *any* direction from here is due south (see Figure 15). We aim the figure in one direction and make straight for the equator (let's not worry precisely how we do this), 'transporting' the vector over the surface of the Earth in a 'straight line' which, as we know, is actually a geodesic. On reaching the equator we turn the chariot towards the east, but of course the gear mechanism ensures that the figure continues to point south. We travel along the equator for about 10,000 kilometres, or one quarter of the Earth's circumference, before turning north again. The figure continues to point south as we make our way back to the North Pole.

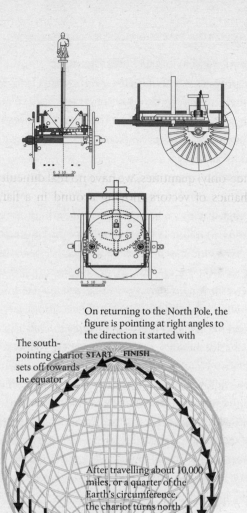

Figure 15. A south-pointing chariot (shown above) sets off from the North Pole and heads for the equator (below). On reaching the equator it turns east for 10,000 miles before heading north again. When it arrives back at the North Pole we find that it is pointing at right angles to its starting direction. This is the parallel transport of a vector quantity.

We return the chariot to its starting point, but the figure now points in a direction which forms a right angle (90°) to the direction it started with. We've done nothing physically to change the orientation of the vector, but the simple process of 'parallel transporting' it around the surface of a sphere has nevertheless changed its direction.

Now, the direction in which a vector is pointing is a critically important feature in any physical theory, so a theory crafted in any kind of curved space, such as general relativity, must not only somehow take the effects of such parallel transport into account, but must also make such effects completely independent of the choice of coordinate system, as demanded by the principle of general covariance. Einstein was well aware of this.

One solution, developed by Italian mathematician Tullio Levi-Civita, relies on the curvature of the space itself.* In our south-pointing chariot analogy, instead of moving the chariot imagine that we're able to rotate the Earth, so that the chariot traces the same geodesic on the surface. This method of parallel transport is called the *Levi-Civita connection on the sphere*. Instead of trying to figure out how to describe the transport of the vector over the surface of the spherical Earth using a potentially complex system of coordinates (x, y, z), we let the geometry (and particularly the symmetry) of the sphere itself do all the work. The connection then provides a natural means of transporting the vector over the surface in a consistent way, without recourse to an unnatural coordinate system.

---

* Levi-Civita was a student of Gregorio Ricci-Curbastro, who invented the tensor calculus that Einstein used to formulate general relativity, and in the period 1915–17 Levi-Civita corresponded with Einstein on problems related to the mathematical structure of the theory. He wrote an introduction to the Italian translation of Einstein's *Relativity: The Special and the General Theory*, first published in 1921.

The idea of connections might seem all rather obscure, but in fact the quantum field theories that form the standard model of particle physics can also be thought of as 'connection theories'.* In these examples it is the quantum field that parallel transports a vector (such as a spinning electron), and if the field is 'curved'—if its strength changes magnitude and direction from one part of the field to the next—then moving a vector around a closed loop may change its direction when it returns to its starting point, just as the figure on the chariot ends up pointing in a different direction when it gets back to the North Pole. It so happens that in quantum field theories we arbitrarily nail the field onto a background flat, Euclidean space, so that we can describe the vector as moving from 'here' to 'there', and back to 'here'.

And there's the big idea. If general relativity can be reformulated as a connection theory, then—perhaps—it might begin to resemble a classical field theory which can then be quantized, much as Maxwell's classical electromagnetic field was quantized to produce QED. However, there's a big difference. In a connection-theory formulation of general relativity, the gravitational field is the system of connections, but, of course, it is no longer necessary to nail this onto an arbitrary background spacetime. The spacetime metric itself *emerges* naturally from the system of connections.

This insight has a long history. In the late 1940s, Einstein (and others working independently in parallel, including Schrödinger) had attempted to unify general relativity and electromagnetism by reformulating the former using a system of Levi-Civita

---

* In fact, in 1918 the German mathematical physicist Hermann Weyl was in pursuit of a connection theory when he discovered the basis for what became known as a 'gauge theory', a specific type of field theory that, broadly speaking, forms the basis for all modern quantum field theories.

connections as the primary variables. But it was tough going, and Einstein was filled with doubts. At one point he remarked to Schrödinger: 'We have squandered a lot of time on this thing, and the result looks like a gift from the devil's grandmother.'[6] These early attempts at reformulating general relativity got bogged down in the mathematics, and nothing came of them.

Now, general relativity formulated in a system of connections would seem to represent a whole new level of abstraction. This really shouldn't come as too much of a surprise. If the idea is to develop a theory from which spacetime itself emerges, then (let's face it) it can hardly be intuitive, easy to grasp, or to visualize. Sorry.

By the late 1970s, a similar logic had been applied to the parallel transport of objects carrying spin angular momentum, leading to systems of connections—called spin connections—that had found some useful applications in theories of solid state physics. The notion of spin connections wasn't new—it had originated in the 1930s—but Sen figured out how it could be used to reformulate the ADM Hamiltonian form of general relativity, replacing the spacetime metric with spin connections from which the more familiar curved spacetime could then be derived. This all looked quite promising—it implied an object with the properties expected of the graviton.

This was really no more than a relatively thin sketch of an idea. Sen acknowledged helpful discussions with a number of theorists, including Charles Misner, and 'suggestions and encouragement' from Indian-born theorist Abhay Ashtekar.[7] Sen had studied for his Ph.D. at the University of Chicago, where Ashtekar was working a postdoc, and they had collaborated on several research papers. Ashtekar was well on his way to becoming recognized as one of the leading researchers on general relativity of his generation.* They had become close friends, enjoying long walks together. Sen would later serve as best man at Ashtekar's wedding in 1986. They remain close friends today.

Ashtekar realized that Sen's idea held the promise of a complete reformulation of the Hamiltonian form of general relativity. Sen left physics and moved into the telecommunications industry, eventually joining Motorola. Over the next couple of years Ashtekar greatly expanded on Sen's ideas and proceeded to work out the consequences. What he discovered was quite remarkable. The switch to spin connections not only greatly simplified the equations, but it also made them look like a classical field theory, perhaps no more than a knight's move from a full quantum field theory.

This is much more than a matter of passing resemblance: it was now 'possible to import into gravity certain powerful mathematical techniques from [quantum field] theories'.[8] This removed the 'wedge' between general relativity and the theory of elementary particles, without sacrificing the emphasis on geometry that is inherent in the former.

Lee Smolin completed his Ph.D. at Harvard in 1979 and spent the next few years working as a postdoctoral researcher at the Institute for Advanced Study at Princeton, the Institute for Theoretical Physics in Santa Barbara, and the University of Chicago. Just as Sydney Coleman had warned, he hadn't made any definitive progress, and was quite puzzled by the fact that he continued to have a career. 'One sure reason was because at the time very few people worked on quantum gravity,' he figured, 'so there was little competition'.[9] At least his colleagues still seemed interested in his work.

---

* Ashtekar's citation at his induction into the US National Academy of Sciences reads: 'Ashtekar made major contributions to general relativity and developed Loop Quantum Gravity to unify General Relativity with Quantum Physics.'

He was appointed as an assistant professor at Yale University in Connecticut in July 1984. He had read Sen's papers when they had first appeared a few years earlier, but his attempts to build on Sen's work didn't immediately bear fruit. He'd known Ashtekar since 1980, and they had remained in contact. When, in 1985, Smolin learned about Ashtekar's work on spin connections during a phone call, he realized that this was just what was needed to transform quantum gravity into a real subject, 'one in which it would in time become possible to do calculations that yielded definite predictions about the structure of space and time on the Planck scale.'[10]

He invited Ashtekar to Yale to give a seminar.

8

OUR SECOND OR
THIRD GUESS SOLVED
THE EQUATIONS
EXACTLY

Abhay Ashtekar had made a significant breakthrough, but its impact was limited to a relatively small group of theorists. There's a simple reason. By late 1984, the first *superstring* revolution was well underway.

String theory has a long history. Its foundations can be traced back to the summer of 1968, when Gabriele Veneziano, a young Italian postdoctoral physicist working at CERN, had puzzled over ways to describe the collision and scattering of particles such as pions, which as we now know contain a quark and antiquark. When a young Yeshiva University professor called Leonard Susskind heard about Veneziano's work, he realized this description was entirely equivalent to what happens when 'two little loops of string come together, join, oscillate a little bit, and then go flying off'.[1]

In Susskind's original description, the 'string' was identified with the 'lines' of colour force which fasten the particles together like a piece of elastic. Similar ideas were developed around the same time independently by Danish physicist Holger Nielsen at the Niels Bohr Institute in Copenhagen, Japan-born American theorist Yoichiro Nambu in Chicago, and Tetsuo Goto at Nihon University in Japan. One thing led to another. What if, instead of treating elementary particles as point particles, we instead treat the elastic that holds them together as the elementary stuff of the universe?

This was a very different kind of big idea, one which—at a stroke—eliminated all the mathematical problems created by treating elementary particles as though all their mass is concentrated to an infinitesimally small point. In a string theory, the different particles are represented by different *vibrational patterns* in a common string type. The mass of a particle would then be just the energy of its string vibration, calculated from $m = E/c^2$.

But early versions of string theory did not look at all promising, and consequently did not attract much attention. It could deal only with strings that described bosons. The theory required as many as twenty-six spacetime dimensions: twenty-five dimensions of space and one of time. The theory also predicted the existence of tachyons, hypothetical particles that possess imaginary mass and travel only at speeds faster than light, causing chaos to the time-organizing principles of cause and effect.

A theory that describes only bosons would seem to be really rather limited, but by now theorists had figured that it was possible to introduce fermions through the 'back door', by—no real surprise—assuming a fundamental spacetime symmetry between fermions and bosons. Theories of supersymmetry were beginning to emerge at around the same time, and some theorists have since argued that supersymmetry was actually derived from early string theory.

It doesn't matter much precisely how it all came about. Combining the big idea of strings with a second big idea of supersymmetry meant that by 1972 some of the more substantial problems of early string theory had been resolved, largely through the work of French theorist Pierre Ramond, working at the National Accelerator Laboratory in Chicago, and by American John Schwarz and French theorist André Neveu at Princeton University.

Schwarz found that the number of spacetime dimensions required for superstrings to vibrate in was no longer the twenty-six demanded by early versions of string theory. It was just ten: nine dimensions of space and one of time. And the tachyons were now gone.

There were still plenty of other problems, and the theory was still not very highly regarded. One measure of the recognition afforded by a scientific contribution is the ease with which the contributor can then get a job or secure tenure at a prestigious university. By this measure, the signs weren't good. Ramond was denied promotion to a tenured professorship at Yale. Schwarz's work on string theory wasn't enough to secure him tenure at Princeton. Instead, he took a research associate position at the California Institute of Technology (Caltech) in Pasadena, and continued to work on superstring theory with another French theorist, Joël Scherk.

Schwarz later explained that they persisted with the theory because they were struck by its mathematical beauty, believing that it must surely be good for something. But there was one other thing:[2]

> So, one of the problems that we had had with the string theory was that in the spectrum of particles that it gave, there was one that had no mass and two units of spin. And this was just one of the things that was wrong for describing strong nuclear forces, because there isn't a particle like that. However, these are exactly the properties one should expect for the quantum of gravity.

Superstrings come in two forms: open and closed. Open strings have loose ends that we can think of as representing charged particles and their antiparticles, one at either end, with the string vibration representing the particle carrying the force between them.* Open strings therefore predict both matter particles *and* the forces between them. But the theory also *demands* closed strings. When a particle and antiparticle annihilate, the two ends of the string may join up to form a closed string.

But if there are closed strings, then there are also particles with no rest mass and two units of spin, which are precisely the properties expected of the graviton. This was a particle that had no place at all in particle physics, and its prediction by the theory was seen as rather problematic. In 1974 Schwarz and Scherk turned it into a virtue. They argued that superstring theory appeared to be saying that *all* nature's forces are just different vibrational patterns in open and closed strings. In superstring theory it seemed that these forces are automatically unified. It promised to be not only a theory of the strong colour force, but potentially a *theory of everything*.

It seemed that all the elementary particles then known—their masses, charges, spins—the forces between them and all the parameters that in the standard model of particle physics couldn't be derived from first principles, could be subsumed into a single theory with just two fundamental constants. These determine the tension of the string and the extent of the coupling between strings.

Admittedly, the nature of the strings themselves had to be reinterpreted. Introducing some aspects of gravitational physics meant that the strings could now no longer be interpreted as bits of elastic joining quarks and antiquarks together, hinting at lengths on the order of a femtometre (10^{-15} metres). They were

---

* Please note that I tend to use the terms 'string' and 'superstring' interchangeably, but you can always assume that the objects being referred to are supersymmetric.

now obliged to exist with much smaller lengths, consistent with the Planck scale, twenty orders of magnitude smaller.

But there remained more than a few stubbornly persistent problems, and nobody was interested.

Whilst working for a few months at CERN in Geneva, Schwarz began collaborating with British physicist Michael Green, based at Queen Mary College in London. Together they explored aspects of three different kinds of superstring theory. These became known as Type I, Type IIA, and Type IIB. All require ten spacetime dimensions but differ in the way that the supersymmetries are applied. Like the Minimal Supersymmetric Standard Model, Type I superstring theory makes use of one supersymmetry, whereas Type II superstring theories use two.

Although their work continued to be ignored by the wider community, the theory was beginning to gain interest and win a few advocates. Among them was Princeton mathematical physicist Edward Witten.

By the early 1980s, Witten was a relatively young, 30-year-old theorist. But he was already a phenomenon. He had had a rather eclectic career. After studying history and linguistics at Brandeis University near Boston he went on to study economics at the University of Wisconsin and embarked on a career in politics.

He worked on George McGovern's 1972 presidential campaign. After McGovern's overwhelming defeat by Richard Nixon, he abandoned politics and moved to Princeton to study mathematics. He migrated to physics shortly afterwards. He studied for his doctorate under David Gross, one of the architects of QCD, securing his Ph.D. in 1976. He was a tenured professor at Princeton just four years later.

Witten was establishing a reputation as a bona fide genius, a modern-day Einstein. He was awarded a MacArthur Foundation 'genius' grant in 1982.

Witten's interest and involvement in superstring theory were in themselves sufficient to draw more attention to it. He quickly persuaded Schwarz and Green that if the theory was to become a viable alternative to the standard model of particle physics and a serious candidate as a theory of everything, then they needed to resolve a few more of its troublesome problems. It was not even possible to apply simple, so-called 'one-loop' radiative corrections without a breakdown in the theory's mathematical consistency. This was something that just had to be sorted out.

Schwarz and Green figured out how to resolve this problem in the summer of 1984, and things then happened very quickly. They published a paper detailing their solution in September. Witten submitted his first paper on superstrings to the same journal later that month. At Princeton, David Gross, Jeffrey Harvey, Emil Martinec, and Ryan Rohm (who would collectively come to be known as the Princeton String Quartet) found yet another version, called heterotic (or hybrid*) superstring theory. It turned out that there are two types of heterotic superstring theory, both requiring ten spacetime dimensions, making five versions of the theory in total. They submitted their paper for publication in November 1984.

The first superstring revolution had begun.[†]

In the early 1980s Lee Smolin was intrigued by superstring theory and was keen to find out more about it. But he recognized that if it was to have more than just pretensions to a theory of quantum gravity (let alone a theory of everything), then something had to be done about the theory's presumption of a

---

* Heterotic superstring theories are hybrids of Type I and bosonic superstring theories.

[†] In the late 1970s and early 1980s, about 50 research papers were published on the subject of superstring theory each year. In 1984 this climbed to about 100. In 1987 1,200 papers on superstring theory were published.

background spacetime. Switching from a description based on point particles to one of strings had eliminated some of the mathematical unpleasantness, and introducing supersymmetry had reduced some of the awkward dimensionality of the theory and got rid of the troublesome tachyons (albeit at the cost of introducing many more parameters and the problem of super-symmetry-breaking). But the superstrings were still imagined to be moving about in nine dimensions of space, and one dimension of time.

Smolin had gained his convictions about the background the hard way, by trying various approaches and failing, sometimes pretty spectacularly. During his time as both a Ph.D. student and postdoctoral researcher, his ideas had been shaped by a series of encounters with some leading particle theorists, such as American Kenneth Wilson and Russian Alexander Polyakov, and a relatively unknown, independent theorist from England called Julian Barbour. These early encounters were to prove enormously influential and would help to shape virtually his entire scientific career.

From Wilson and Polyakov, Smolin learned about the possi-bilities afforded by reducing continuous spacetime to a regular *lattice* of fixed points and distances. This approach had become necessary in order to perform calculations using QCD.

Although the equations of QCD can be written down in a rela-tively straightforward manner, they cannot be solved analytically, 'on paper'. The colour force is extremely strong, and so the cor-responding energies of colour-force interactions are therefore very high. And because the gluons that act to bind quarks together also carry colour charge (red, green, blue), everything interacts with everything else. Virtually anything can happen and keeping track of all the possible virtual and elementary-particle permuta-tions in QCD is extremely demanding.

·Also renormalization, of the kind applied through small radiative corrections (or 'perturbations') based on Feynman diagrams used so successfully in QED, simply doesn't work for QCD. It can be applied in short-distance interactions where the colour force coupling is small (remember, this force works in the opposite sense to electromagnetism). But for most circumstances of interest, theorists have no choice but to solve the equations on a computer.

In lattice QCD, the magnitudes of the quantum fields representing the quarks are defined only at specific points on a three-dimensional grid or lattice (rather than continuously, as would be required for a continuous field). The magnitudes of the gluon fields are then defined on the links connecting neighbouring points on the lattice.

Quarks are therefore imagined to live only on the lattice points, and the 'lines of force' representing the gluons live only along the links joining neighbouring points together. If it helps, you can think of these lines of force in much the same way that you think about the magnetic field surrounding a small bar magnet, pictured in Figure 1 in Chapter 1. Of course, it doesn't require a great leap of imagination to then think of these lines as 'strings'.

For the purposes of calculation, the distance between lattice points is assumed to be on the order of a few tenths to a few hundredths of a femtometre (10^{-15} metres), and the more this is reduced, the closer we get to a 'continuum'. To reduce the number of computer calculations required, the theorists perform calculations at smaller and smaller distances and then extrapolate their results to zero. This technique continues to be used to this day. Great accuracy is possible but only at a cost; the most rigorous lattice QCD calculations require the world's largest supercomputers.[3]

In lattice QCD, the lattice is simply a mathematical device for making the calculations accessible. But if we take it seriously, it imposes a 'quantization' of space into a system of discrete points and distances. The distance between points doesn't necessarily have to be extrapolated to zero. There could be a stubbornly finite, but extremely small, distance between the points, representing an ultimate distance that cannot be shortened. Perhaps something on the order of the Planck length.

Both Wilson and (independently) Polyakov had hoped that it might be possible to formulate QCD in a way that would render it analytically soluble, by focusing on the properties and behaviour of the 'lines' themselves, which became known as 'Wilson lines' or 'Wilson loops', essentially quantized lines or loops of colour force.

In quantum field theory, we typically think of the field as primary, with lines of force or lines of constant field strength used to 'map' the field, much like the contour lines on a two-dimensional geographical map. We realize that what look like discrete lines in our experiment with iron filings is really an illusion—the magnetic field is continuous and the lines are an artefact created by the visibly finite sizes of the filings. But there are genuine examples of quantized lines of force in physics, for example in the magnetic properties of superconductors. Wilson and Polyakov proposed to make the lines primary and the field secondary, derived from the lines. And this, of course, is precisely what string theory does.

Another characteristic of the lines of force is that they are imagined to *flow*, from north to south (magnetic field) or positive to negative (electric field), or from one type of quark colour to another (colour field). But if we remove the magnetic poles and the electric charges and the quarks (in other words, if we remove all the matter), then we can imagine that the field lines do not cease to exist but instead form *closed loops* of force, analogous in many ways to the closed strings of string theory. The perspective then becomes one of lines and loops of force.

Polyakov went further. He attempted to formulate QCD entirely in terms of the lines or loops, without the need for a background lattice.

Smolin had tried to apply these ideas to gravity in his first research paper as a postgraduate student, but he could only get so far. A fixed lattice represents exactly the kind of background spacetime that general relativity eliminates so successfully, and the structures are incompatible. 'The key lesson I learned from this failed attempt was that one cannot fashion a successful quantum theory of gravity out of objects moving against a fixed background.'[4] He needed a lattice that could become dynamic, one that would get more directly involved in the physics, as spacetime itself does in general relativity. He learned from Barbour that the theory he was searching for could only be understood as a dynamically evolving network of relationships, from which our more familiar spacetime somehow emerges.

When Schwarz and Scherk revealed that superstring theory included the graviton and promised the possibility of a quantum theory of gravity, Smolin became determined to find a way to make the theory background-independent, such that conventional spacetime would emerge as a limit or approximation. Whilst working as a postdoctoral researcher at the University of Chicago, he reached for the methods he had learned from Wilson and Polyakov, enlisting the help of a graduate student, Louis Crane.

Crane was no ordinary student. He was a mathematical prodigy. He had enrolled at the university as a young teenager, only to be expelled for his political activism. It had taken him ten years to find a way back.

Together they tried two distinct approaches, one of which produced a couple of published papers. Their efforts helped to convince Smolin that the theory had somehow to be based on the relationships established in an evolving network of Wilson lines, or loops. He just wasn't sure how this might be achieved.

Smolin invited Ashtekar to give a seminar on his spin-connection reformulation at Yale in December 1985. He quickly became convinced that this was precisely what he had been looking for. Having failed to drive Ashtekar back to the airport in Hartford,* on his return to the office he sat down to apply the new approach to the structure he had developed with Crane in Chicago.

He had already arranged to spend six months at a programme titled 'Rapprochement of Approaches to Quantum Gravity', coordinated by Ashtekar, David Boulware, and Ted Newman at the Institute for Theoretical Physics in Santa Barbara,† which began in January 1986. He recruited support from two friends, graduate student Paul Renteln (who had also sat attentively through Ashtekar's seminar at Yale) and Ted Jacobson, a Research Associate in the Department of Physics at the University of California, Santa Barbara.

Working with Renteln, Smolin now applied an approach based on loops of gravitational force to the spin-connection formulation developed by Ashtekar. The resulting theory was one of loops only, as gravitational force doesn't 'flow' from one pole or charge to another, and so appears only in the form of closed loops (just as it is ascribed to closed strings in string theory). At first sight, this seems to contradict the assertion (and the title I've chosen for Chapter 2) that there is no such thing as the 'force' of gravity. But this assertion is based on the idea that spacetime

---

* Smolin suffered two flat tyres and had to leave Ashtekar to hitch a lift to the airport while he waited at the roadside for help. Ashtekar made his flight, however.

† This became the Kavli Institute for Theoretical Physics in the early 2000s, in recognition of a $7.5 million donation by the Norwegian-American entrepreneur Fred Kavli.

itself is the gravitational field. In the formulation that Smolin and Renteln were now attempting, the logic is reversed: the idea of gravitational force becomes primary, with spacetime (the gravitational field) emerging as a secondary phenomenon.

The structure that Smolin and Renteln obtained suggested some very simple rules for how the loops interact on the lattice. But, of course, to make any real progress it would be necessary to get rid of the lattice entirely and deal only with the relationships between the loops, as Polyakov had argued.

And so, one day in February 1986, Smolin found himself standing alongside Jacobson in front of the blackboard in a small lecture room at the institute. Together they had recast the equations so that they referred only to the loops and the gravitational field they carried around with them, not the lattice. Emboldened, they had guessed at a possible analytical solution for these equations. Just as Schrödinger's wavefunctions describe the quantum states of the electron in a hydrogen atom, solutions to these equations describe the quantized loops of gravitational force, or the *quantum states of the geometry of spacetime*.

What they found came as a great surprise. More than a decade of frustration had taught the theorists to be wary and expect the worst. Smolin and Jacobson had duly limited their ambition to find only approximate solutions. But they now found that their second or third guess solved the equations *exactly*. As Smolin later recalled:[5]

> I still remember vividly the blackboard, and that it was sunny and Ted was wearing a T-shirt (then again, it is always sunny in Santa Barbara and Ted always wears a T-shirt).

Within a few days they had worked out that there is an infinite number of such exact solutions, provided they apply some simple rules governing the intersection of the loops.

Although it would not become obvious for some time, these proved to be formal solutions only, and as such they hold no physical significance (they imply zero volume). There was much more still to be done. But it seemed that the switch to a spin-connection formalism had unlocked an important clue. It suggested the promise of a theory which has no need of a background spacetime framework. It promised a theory not of loops existing *in* space. Rather, the theorists imagined that the relationships between the loops would *define* space. This was still a big step in the right direction.

It's worth pausing for a moment to reflect on this. Ashtekar had deduced a version of general relativity which closely resembles a classical field theory. Smolin, Renteln, and Jacobson had now quantized this with the aid of Wilson loops and, although there were still many unanswered questions, these early results were extremely encouraging. It had not been necessary to introduce any other new concept, such as supersymmetry, that was likely to proliferate untold numbers of adjustable parameters or problems associated with supersymmetry-breaking. All that had been required was the introduction of some new variables—which became known as *Ashtekar variables*—based on the idea of spin connections.

There was lots to do but the next step was to demonstrate that these solutions are genuinely independent of the background, by showing that they are unaffected by (or 'invariant' to) any arbitrary change of coordinate system. Smolin figured that they had already done the difficult bit. Showing that the solutions are invariant should have been quite straightforward.

Smolin and Jacobson presented their results at a conference organized towards the end of the Santa Barbara workshop. In the audience was a young Italian theorist who had just completed his Ph.D. at the University of Padua. His name was Carlo Rovelli.

Now armed with a freshly minted Ph.D., in early 1986 Rovelli had scrambled to pull enough money together from various sources (including his own savings) to fund some trips abroad. He had a singular goal in mind. He wanted to meet 'the greatest figures in the world of quantum gravity'.[6]

His first stop was Imperial College, London, and Chris Isham, the author of the review article that had opened his eyes to the problems that beset the subject. He poured out his confused ideas and listened intently to what Isham had to say about them, mulling over his words during long walks alone in nearby Kensington Gardens. He made photocopies of every paper he could lay his hands on and read voraciously. He stayed in London for two months.

From Isham he learned about the promising but still unpublished work of a young Indian theorist called Abhay Ashtekar at Syracuse University in New York.* He learned more from Ashtekar's own handwritten notes on the new variables, which were now circulating within the small community of theorists (Rovelli remembers that some copies had been printed on violet-coloured paper). He promptly wrote to Ashtekar and asked to visit, spending a further two months at Syracuse learning about the new reformulation of the ADM Hamiltonian form of general relativity directly from its author.

Although unknown to the theoretical physics community in America, he used his own money to finance another trip to attend the conference in Santa Barbara. He simply turned up, uninvited.

---

* Ashtekar submitted a paper titled 'New Variables for Classical and Quantum Gravity' to the journal *Physical Review Letters* in December 1985, but this was not published until November 1986. Ashtekar admits this was entirely his own fault—his referees had requested a few minor changes, 'but because the Santa Barbara workshop was so exciting and so many new results were coming out, I delayed my response by over 6 months!' (Abhay Ashtekar, personal communication, 11 December 2017).

He sat in the audience and listened as Smolin and Jacobson explained how they had applied an approach based on Wilson loops to discover some exact solutions of Ashtekar's formulation.

He didn't talk much with Smolin at the conference, but he already knew where he wanted to go next.

In the meantime, the superstring theory programme had taken a decidedly unpromising turn.

Introducing two big ideas—strings and supersymmetry—had helped to resolve some of the mathematical problems with the theory, but the physicists had been left to work out what to do with the fact that they had uncovered *five* different types of superstring theory, all seemingly equally valid. They also had to decide what to do about the six 'extra' spatial dimensions that all versions of the theory demanded. Our experience is that space is determinedly three-dimensional, so the string theorists had to work out how to roll up the extra dimensions and hide them out of sight.

In 1984, American theorist Andrew Strominger, then at the Institute for Advanced Study in Princeton, searched for a way to do this in collaboration with British mathematical physicist Philip Candelas, then at the University of Texas. His search led to the library, and a recent paper by Chinese-born American mathematician Shing-Tung Yau. The paper contained a proof of something called the Calabi conjecture, named for Italian-American mathematician Eugenio Calabi. The proof confirmed the existence of a series of shapes—now called Calabi–Yau shapes or spaces—that appeared to be just what was needed to hide away the extra dimensions. An example is shown in Figure 16.

Strominger and Candelas got in touch with Gary Horowitz, at the University of California in Santa Barbara, a physicist who had worked with Yau as a post-doctoral research associate. Strominger also visited Witten, to discover that the latter had independently arrived at the same solution.

Figure 16. Calabi–Yau shapes are complex, higher-dimensional algebraic surfaces used in superstring theory to 'hide' the six extra spatial dimensions required.

The four theorists collaborated on a paper which was published in 1985. Thus was born the idea of 'hidden dimensions'. If I could mark an infinitesimally small point on the desk in front of my keyboard, and could somehow zoom in on this point and magnify it so that a distance of a billionth of a trillionth of a trillionth of a centimetre becomes visible, then superstring theory says that I should perceive six further spatial dimensions, curled up into a Calabi–Yau shape.

Although the theory required that these dimensions be hidden, this didn't mean they have no influence on the physics. Quite the contrary. The precise shape of the Calabi–Yau space, and especially the number of 'holes' it has, determines the nature of the superstring vibrations that are possible. It thus determines the physical constants, the laws of physics, and the range or spectrum of particles that will prevail. In other words, the shape determines what kind of physical universe we get. It seemed that the theorists just needed to find the specific Calabi–Yau shape that is consistent with the particles of the standard model and the laws of physics governing their properties and behaviour. They needed to find a Calabi–Yau shape consistent with our universe.

But, as so often happens, nature wasn't prepared to play nicely. In *The Hidden Reality*, Brian Greene wrote of his early experiences working with these mathematical shapes:[7]

> When I started working on string theory, back in the mid-1980s, there were only a handful of known Calabi-Yau shapes, so one could imagine studying each, looking for a match to known physics. My doctoral dissertation was one of the earliest steps in this direction. A few years later, when I was a postdoctoral fellow (working for the Yau of Calabi-Yau), the number of Calabi-Yau shapes had grown to a few thousand, which presented more of a challenge to exhaustive analysis—but that's what graduate students are for. As time passed, however, the pages of the Calabi-Yau catalog continued to multiply... they have now grown more numerous than grains of sand on a beach. Every beach.

Worse still, assuming that such a shape really does exist at the margins of the three-dimensional space with which we are familiar, it seemed that there was absolutely no way to identify the unique shape that is consistent with our universe. In a paper published in September 1986, Strominger had to admit: 'All predictive power seems to have been lost'.[8]

There were, in any case, some rumblings of dissent, particularly from the older generation of theorists. In a book published just a few years later, Feynman criticized the string theorists for 'not calculating anything' and 'not checking their ideas'.[9] Glashow argued that string theorists 'cannot even be sure that their formalism includes a description of such things as protons and electrons'.[10]

Some did become discouraged and left the field altogether. But many remained, stubbornly refusing to believe that string theory was dead. The bandwagon, such as it was, was already rolling, and careers were being built (Schwarz was appointed as Professor of Theoretical Physics at Caltech in 1985). In one sense, there was now too much at stake.

9

I USED EVERY
AVAILABLE KEY RING
IN VERONA

Lee Smolin and Ted Jacobson submitted a paper on their breakthrough to the journal *Nuclear Physics B* in August 1987, by which time Jacobson had taken a postdoctoral position at Brandeis University near Boston, Massachusetts, working with Stanley Deser. Despite their success, it was still not possible to determine whether 'quantum general relativity may or may not be a sensible theory', but the next steps seemed clear enough.[1]

Reading through this paper again for the first time in decades, Jacobson expressed mixed feelings. It 'reminded me both of how much fun and excitement we felt, as well as why, by the end of that project, I had pretty much concluded that this was not in my view a very promising avenue'.[2] Jacobson turned his attention to the physics of black holes, hoping to build a more physical, bottom-up approach to quantum gravity.

Although the exact solutions that Smolin and Jacobson had deduced were later shown to be unphysical, Smolin believed they had glimpsed the first few quantum states of the geometry of space. This was supposed to be the difficult bit. The next task was to solve a second set of equations that would demonstrate that these quantum states are entirely independent of the choice of coordinate system or, in other words, that they are generally covariant. This was supposed to be the easy bit.

But their early optimism was quickly dashed. Back at Yale and working again with Crane, who had now been appointed as an assistant professor, Smolin discovered that they could find quantum states that solved one set of equations or the other, but they could find none that solved both together. They became convinced that what they were trying to do was actually impossible.

Carlo Rovelli had elected to study physics at university partly as a means to avoid obligatory national service. With his Ph.D. completed, the call to service now loomed, ominously. He refused, expecting to serve a two-year prison sentence which he figured was still preferable to one year of military service. But he got lucky, and was detained for only a few days: 'in those days the Italian government was sending soldiers abroad—to Lebanon—for the first time since the Second World War, without real support from the public. So they did not want people like me making waves and decided to let me go.'[3] On his release, he returned to Syracuse and engaged in some further work with Ashtekar, and made plans to visit Lee Smolin at Yale.

Whilst in Syracuse, he received some bad news. His long-term Italian girlfriend had decided to end their relationship. Rovelli was very much in love with her, and although they eschewed such old-fashioned notions as marriage (and they regarded the idea of 'engagement' as belonging to an older century), he had imagined that they would nevertheless be life-partners. He was utterly devastated. He was due to leave for Yale the next day, but was now in no mood to go. He almost cancelled the trip.

After some soul-searching, he decided that this was a visit that he really couldn't postpone, and so it was that he arrived in Smolin's office one day in October 1987. Their discussion was rather difficult. 'I explained to him that there was nothing to do, because we were completely stuck,' said Smolin.[4] Smolin went on to suggest that although Rovelli was quite welcome to stay, as the task they were intending to work on seemed to be impossible, perhaps he might want to consider returning to Italy.

This was too much. Rovelli's relationship had ended bitterly, and he was in the 'darkest desperation'.[5] It now seemed that his scientific ambitions were to be frustrated, too. Unable to control his emotional state, he became tearful. Seeing that Smolin was rather startled by this behaviour, Rovelli quickly apologized and explained the reasons for his distress. Smolin was sympathetic. He confessed that he, too, had recently experienced a similarly tragic end to a relationship.

There was an awkward silence.

In an attempt to change the subject and lighten the mood, Smolin asked if Rovelli liked to sail. Rovelli smiled. He was a keen sailor. They promptly left the office and headed for the Yale University boat house. They took a small dinghy and spent the rest of the day sailing around a small bay eastward along the Connecticut coastline. They forgot all about their scientific problems and talked instead about their lives, dreams, and troubled relationships.

Rovelli rented a room in Crane's apartment, and Smolin saw nothing of him the next day. But the day afterwards, Rovelli appeared at the door to Smolin's office.

'I've found the answer to all the problems,' he said.[6]

In the version of the theory that Smolin, Jacobson, and Crane had wrestled with, they had rid the structure of the lattice but were expressing the quantum states as functions of the spin connections. But the spin connections still carry the gravitational field around with them, and, of course, in general relativity the gravitational field *is* spacetime. So, despite how it seemed, even without the lattice the background spacetime had still not been completely removed. This was why the solutions to the equations were failing the coordinate system test.

Rovelli realized that by applying a further reformulation, using techniques he had learned from Isham, they could express the quantum states in terms of the loops of force only. The same logic would still apply. This would be a theory of intersecting loops, or 'loops of gravity'. Such loops would be imagined not to exist *in* space: collectively, they *are* space.

Within a day they had convinced themselves that this was a good idea. It took a few months to work out the detail:[7]

> In the end we had a theory of the kind that Polyakov had spoken about as his great dream: a theory of pure loops which described an aspect of the real world in equations so simple they could be solved exactly. And when it was used to construct the quantum version of Einstein's theory of gravity, the theory depended only on the relationships of the loops to one another—on how they knot, link and kink.

They presented what they had found to Ashtekar in Syracuse. On the way to the airport Smolin's Dodge Dart was shunted from behind by a guy in a Maserati. In the brief and rather unequal battle between American steel and Italian style, steel won hands down. The Dodge escaped with barely a scratch, but the Maserati was a write-off.

Rovelli had developed a high fever, but nevertheless stood to present their results. There followed a long silence, broken by Ashtekar who commented that this was the first time he had ever witnessed a set of equations that might just be the quantum theory of gravity.

They flew to London shortly afterwards, and presented their results to Isham before setting off for an international conference on gravity and cosmology held in Goa, India, on 14–19 December 1987. Rovelli had no invitation to the conference, which was now closed, but on an impulse he decided to trust his luck and go anyway. His audacity was rewarded with the best room in the hotel where conference participants were staying. Ashtekar recalls that whilst waiting to check in at the airport for his return flight, Rovelli struck up a conversation with a rather beautiful young woman. She was taking the same Air India flight but travelling first class. At the gate he enquired if it might be possible for him to get an upgrade, so that they could continue their conversation. 'And, being Carlo, it worked!'[8] Or not. Although now both seated in first class, the cabin layout precluded intimate conversation and she paid him no attention.[9]

Their presentation was well received. The 'loop-space representation of quantum general relativity' became more popularly known as *loop quantum gravity*.*

This was not the end, but the beginning. Years of hard work now beckoned to the small community of theorists who would dedicate their scientific careers to the elaboration of loop quantum gravity, or LQG. It was one thing to establish a set of equations that could be solved exactly. Understanding what the solutions actually meant was quite another. As Smolin later admitted:[10]

> In my experience it really is true that as a scientist one has only a few good ideas. They are few and far between, and come only after many years of preparation. What is worse, having had a good idea one is condemned to years of hard work developing it.

* Rovelli coined this name in a review published in 1998, and it took many years to catch on. Other theorists suggested different names. Alejandro Corichi recalls bar-discussions dating to 1996, when the theory was being generally referred to as the 'loop representation of non-perturbative quantum gravity' (personal communication, 21 November 2017).

It would, in fact, take about *eight* years of hard work to get from this initial breakthrough to the result that would deliver a real understanding of the quantum nature of space and serve to define what LQG is all about. The journey would involve three distinct steps.

From loops to knots

Where to start? This was a theory based on the idea of loops of gravitational force, so a potentially fertile area for exploration of the properties of the loops—the 'quantum states of space'—was the theory of *knots*. Now, knots are very familiar to anyone who has ever tied shoelaces or ropes or worn a tie, and the notion that there exists a substantial body of mathematical theory devoted to knots might seem rather incongruous. But the mathematician's knot is a little different from the knots we encounter in everyday life. For one thing, these are knots in objects which form continuous loops, rather than knots created in a single one-dimensional string (such as a tie) or by tying together the ends of individual strings (such as shoelaces).

A closed loop with no 'crossings' or links is called the *unknot* and, rather obviously, does not contain any knots. One of the simplest examples of a knot is the *trefoil*, shown in Figure 17. Such objects are three-dimensional but to visualize and manipulate them 'on paper' it is necessary to project them onto a two-dimensional plane, by showing where those parts of the loop (the 'overstrands') lie above or 'cross' other parts (the 'understrands', usually depicted as breaks in the loop). If you spend long enough studying a trefoil knot you will be able to convince yourself that it is indeed knotted—it can't be transformed by any kind of smooth manipulation into the unknot. To unknot it we must break the knot and rejoin the ends.

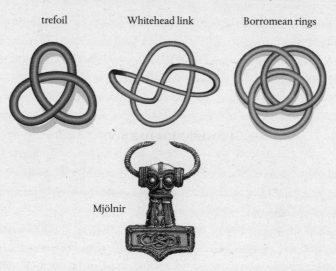

trefoil   Whitehead link   Borromean rings

Mjölnir

Figure 17. Examples of knots include the trefoil, the Whitehead link (which features on this depiction of Thor's legendary hammer Mjölnir), and the Borromean rings.

Things become a little more complicated if we have two or more loops. Two loops can be linked, such that they cannot be separated without breaking one of the loops. In the rather devious *Whitehead link* (named for British mathematician J. H. C. Whitehead) one unknot is threaded through a second unknot which is twisted. Again, the loops can't be separated. The result appears on some depictions of Thor's legendary hammer, Mjölnir.* Three loops can be linked to form the *Borromean rings*. These might look like a set of puzzle rings but don't be fooled:

---

* Decorating Thor's hammer with representations of knots is a tradition that continues to this day. Observant fans of the movie series that defines the Marvel Cinematic Universe may have noticed that Thor's hammer is decorated with a simple trefoil knot.

although no pair of loops are linked, all three together cannot be separated.

Knots obviously have a long history in human culture, as the reference to Thor's hammer demonstrates. The early knot theorists were interested in establishing a mathematical language that could be used to determine whether a complex loop with lots of twists is actually knotted, or a combination of loops is actually inseparably linked.* They also sought to catalogue all the different possible types of knots and links (more than six billion are known today) and to establish mathematical relationships between them.

Knot theory also has a relatively long history of applications to physics, dating back at least to the 1860s. In Lord Kelvin's vortex theory of the atom, atoms were imagined to be formed from knots and links in the ether. In a paper published in 1869, Kelvin depicted a series of knots and links which include the trefoil, linked rings, and Borromean rings. Today knot theory has wide-ranging applications in physical science, including topological quantum field theory and the physics of complex biomolecules such as DNA and proteins although, for reasons that will soon become apparent, it is now of much less relevance to LQG.

The relationship between mathematicians and theoretical physicists can be surprisingly tricky. An innocent bystander might imagine that mathematicians and theoretical physicists are in constant communication, routinely trading insights on problems in their disciplines and potential solutions. But the simple truth is that these two communities don't actually communicate with each other very often. As Smolin started to take an interest

---

* Anyone who has spent frustrating minutes trying to untangle the mess of cables behind the TV, set-top box, DVD player, and sound system will have pondered on this kind of problem.

in knot theory, Crane had to remind him that he had already met one of the world's leading knot theorists in Chicago—Louis Kauffman. Smolin's girlfriend had attended a series of lectures delivered by Kauffman on knots and art at the Art Institute.

Smolin and Rovelli made some progress using knot theory to classify the different possible quantum state solutions, but the physical meaning of these states remained rather obscure. They faced two problems.

Ever since Schrödinger's breakthrough with the elucidation of the quantum states of the electron in a hydrogen atom in 1925, the language of quantum mechanics had drawn attention to the notion of the 'observables' of the theory, and their corresponding mathematical 'operators'. The observables are the things that we literally 'see' or, more correctly, measure when we perform an experiment. They are the properties we then ascribe to the object under study, such as energy, momentum, and spin.

In quantum theory, each observable is related to a corresponding operator, which is simply a mathematical instruction which is applied to the wavefunction, such as multiply it or differentiate it. What happens in quantum mechanics is that when we apply the operator for energy (for example), out pops the discrete, quantized *value* of the energy of the wavefunction—the observable.

General relativity is a classical theory and doesn't rely on operators as such.* It describes the relationship between mass-energy and spacetime, and if we take all the matter away there is, in principle,… well what, exactly? This was the first problem. The loop quantum gravity description was meant to be a description from which *space itself* emerges. But, reformulated as a quantum

---

* The classical 'observables' associated with general relativity are non-local and extremely complex. It is therefore extremely difficult to derive the corresponding quantum-mechanical operators from these.

field theory, in the absence of matter the operators and observables were simply not defined. There did not seem to be an operator for 'space' and therefore no corresponding observable.

The second problem concerned time. As Dirac had discovered, in a constrained Hamiltonian reformulation of general relativity, time 'disappears'. This result is carried through to the Wheeler–DeWitt equation, and the ADM formulation. Everything that Sen, Ashtekar, Smolin, Jacobson, Crane, and Rovelli had done to further reformulate the theory had *not* resulted in the sudden reappearance of a time dimension.

These problems meant that although solutions could be found and classified using knot theory, there was still no telling how they should be interpreted. Rovelli and Smolin had to admit that 'the physical interpretation of the physical loop states that we have found is an open problem'.[11]

Smolin had by this time moved from Yale to Syracuse, allowing him to work more closely with Ashtekar. The pair was a study in contrasts. Ashtekar was rational, analytical, and methodical, paying close attention to the detail of a problem before declaring that he had found a solution. His personality was reflected in the state of his office, which was 'a scientific monastery. There are no stray papers in sight; tape dispenser, stapler, and pencil holder line up with regimental order on the desk.'[12]

Smolin was much more impetuous, full of restless energy, often sacrificing the detail (which would come later) in search of a creative short-cut to the solution. Smolin's office, just three doors down the corridor, looked as though it had been caught in a hurricane, with books, journals, and clothes littering every available surface. He explained: 'I don't really use this room much as an office, more like a closet.'[13]

Rovelli had secured a postdoctoral scholarship from the Italian National Institute for Nuclear Physics (INFN), which allowed him some freedom to choose where to work. He chose the University of Rome, then the home of some of the greatest contemporary Italian theoretical physicists, such as Gianni Jona-Lasinio, Giorgio Parisi, Nicola Cabibbo, and Luciano Maiani. He spent the next few months in the Department of Physics, absorbed with the problems of LQG, working alone at a table in the basement.

But Rovelli's scholarship in Rome lasted only a year, and when it ended he failed to secure any further funding. Cabibbo, then director of the INFN, had heard about Rovelli's work with Smolin and tried to obtain financing for a new contract, but nothing came of his efforts.

These were rather desperate times. Rovelli scrimped and saved, and with some financial support from his father he was able to continue his scientific work for a while. But his prospects were poor. He was working on a subject in which almost nobody in Italy was interested. The warnings of his teachers at the University of Bologna now rang in his ears as he contemplated the end of his career as a theoretical physicist. 'There were moments of distress,' he later wrote.[14]

Then one day in 1989, he received a telephone call. Would he be interested in an assistant professorship at the University of Pittsburgh? His initial reaction was to reject this offer, reluctant to exchange the vitality of a busy European capital with what he perceived to be the soullessness of a dull and sleepy anonymous American town. But he was quickly persuaded by a close friend that he had been gifted a fantastic opportunity to pursue the science that had fascinated him for so many years. Better to be a professional scientist in America than jobless in Italy. He accepted the offer and moved to Pittsburgh in 1990.

If Smolin and Ashtekar were contrasts, then Rovelli brought balance to the group. 'The way each of us organises our thoughts is incredibly different, which can be frustrating,' Rovelli explained, 'Yet we understand together what we couldn't understand separately.'[15]

From knots to weaves

The next significant breakthrough came in late 1990 or early 1991, as Smolin scribbled equations in a notebook, waiting in a noisy garage as mechanics worked to repair his car. He figured out how they might construct and calculate an operator to represent physical area, so that the resulting 'spectrum'—the various quantized values that area could take—would come out finite, without the need for renormalization.

This was the culmination of months of effort developing some novel mathematical techniques, based on aspects of QCD and adapted for operators that are background-independent. Smolin summarized this work in series of lectures delivered to an international conference on theoretical physics in Catalonia, Spain, in June 1991.[16]

In the summer months of 1988 and 1989, Smolin had joined Rovelli in Italy for a couple of annual 'working holidays'. Despite the latter's move to Pittsburgh in 1990 the theorists continued this arrangement for another seven years. The reasons were both practical and personal. Rovelli's father had originally suggested that Smolin stay in a studio apartment next door to his own, and Rovelli had introduced him to a female friend from Verona to whom Smolin would soon become engaged. These experiences 'changed my life as well as my science,' said Smolin. 'LQG owes a lot to the cafes of Piazza Erbe and Piazza Dante [in Verona], where much of our work was done in those days.'[17]

Smolin returned to Italy as a Visiting Scientist at Trento University in the summer of 1991, just an hour or so from Verona by car or train. Ashtekar joined them, and together they worked to apply the techniques that Smolin had developed.

They made some progress. They developed a structure to describe how linked loops of gravitational force could combine to produce an extended 'weave'.[18] This implied once again that our experience of a continuous space, in which objects move smoothly and uniformly, is an illusion of our macroscopic, classical world, just as a smoothly continuous linen sheet is an illusion created by its finely interwoven threads of flax or cotton. Despite appearances, the sheet does not 'exist' between the threads that form it.

LQG suggested that at the Planck scale, space is discrete, composed of individual units or quanta—the loops themselves. These represent the building blocks of space, which is formed from a weave, but more like chain mail produced by linking individual loops of steel than linen produced by weaving continuous threads. Rovelli built a physical, three-dimensional model of such a 'loop space' by threading together as many key rings as he could lay his hands on, shown in Figure 18. He later joked that he had used every available key ring in Verona.

This kind of model suggests that space cannot be continuously variable. At the Planck scale there must be some kind of ultimate area or ultimate volume which cannot be transcended. There can be no area smaller than the smallest area; no volume smaller than the smallest volume. The more familiar areas and volumes of everyday experience must then be derived from these. That these quantities seem to be continuously variable simply reflects the fact that the ultimate units are very, very small, just as the light beam from a torch seems continuous though we know it consists of discrete photons.

The challenge now was to use Smolin's area operator to show precisely how the theory would predict this.

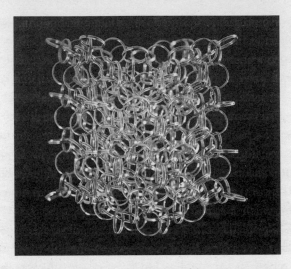

Figure 18. To illustrate how a continuous space could be woven from a network of interlinked loops of gravitational force, Rovelli used 'every available key ring in Verona' to build this three-dimensional model.

From weaves to spin networks

Ashtekar and Smolin visited Trento again in the summer of 1992 on a joint US–Italy cooperative research grant. During a further visit to Verona in August 1993, Smolin and Rovelli took another important step forward. They established how a finite Hamiltonian could be constructed from the background-independent operators for *both* area and volume.

In 1993, Ashtekar was offered the Eberly Chair in Physics at Pennsylvania State University (Penn State), and an opportunity to create a new Center for Gravitational Physics and Geometry. He negotiated to bring Smolin from Syracuse and created another faculty position that was offered to Argentinian theorist Jorge Pullin.

Smolin continued the summer tradition in 1994, taking advantage of a Visiting Lectureship at SISSA, the International School for Advanced Studies based in Trieste, Italy, about 2½ hours from Verona by car. He and Rovelli would work alone in the mornings and would then meet at a café in Piazza Erbe to compare their results. They struggled to make sense of the volume operator, which proved to be much trickier to apply than the area operator. As Rovelli explained: 'It was hard, it was fun, it was great.'[19]

In what was to prove to be the 'final push', they solved the problem, and their perspective changed once again.

Their calculations now suggested that the linking of the loops is rather less important than the network of connections that result (thus rendering Rovelli's key ring model obsolete). Instead of thinking about the loops intersecting, with an overstrand and understand where the loops cross, in this new description the *points of intersection* themselves become much more significant. These become 'nodes' in the resulting network. Applying the volume operator then yields the observable: a specific number of quanta of volume sitting *at the nodes*.

But what of the lines of force connecting the nodes? These now represent the surfaces between adjacent nodes where the volumes of space touch and can be characterized as *areas*. Applying the corresponding operator for area yields a spectrum of areas, related to the square of the Planck length, an unimaginably small 3×10^{-70} square metres.

The networks are therefore characterized by two sets of numbers. These are the values of the volume observables at the nodes and the quantum numbers of the area observables along the links between the nodes. It turns out that the area quantum number (symbol j) can take integral and half-integral values, such as 0, ½, 1, ³⁄₂, 2, ⁵⁄₂.[20]

Smolin realized he had seen these networks before.

I explained at the beginning of Chapter 7 that theorists never start with a blank sheet of paper or an empty drawing board. But there are some who have built entire careers out of their ability to give free rein to their imaginations, by leaving only the barest minimum of necessary physics on the paper, or the drawing board, and clearing off the rest. Such theorists tend to come at the problem from a tangent, or from 'left field'. In the business world, this is sometimes called 'thinking outside the box'.

One such theorist is Oxford mathematical physicist Roger Penrose.

Penrose is not always motivated by a need to respond to the challenges posed by puzzling empirical facts. Sensing that an accepted physical theory does violence to the mathematics or to a cherished philosophical position, or that it simply doesn't 'feel' right, is more than enough. And this, of course, offers the prospects of a *third* road to a quantum theory of gravity, one that doesn't start from general relativity or quantum theory, but in some important sense starts almost from scratch.

Mach (and subsequently Einstein) had tried to reduce all of physics to a system of relationships and interactions between physical objects, in a prescription that doesn't rely on the assumption of a background spacetime. But quantum mechanics not only assumes such a background it further demands that spacetime be regarded as smoothly and continuously variable. The energy and linear and angular momentum of the microscopic objects inside the atom, such as electrons and quarks, might be quantized, but the space they move in is assumed to be continuous and smooth.

This doesn't feel right.

In 1971 Penrose asked himself a seemingly simple and innocent question. Is it possible to replace continuous spacetime with a system built only from the relationships between objects?[21]

My own particular goal had been to try to describe physics in terms of discrete combinatorial quantities, since I had, at that time, been rather strongly of the view that physics and spacetime structure should be based, at root, on *discreteness*, rather than continuity. A companion motivation was a form of Mach's principle, whereby the notion of space itself would be a *derived* one, and not initially present in the scheme.

Penrose's big idea was to build a structure from primary concepts that are already quantized, in such a way that continuous concepts would then emerge as a limit, or an approximation. But which concepts? Penrose had no hesitation in reaching for what he regarded as 'the most quantum-mechanical of all physical quantities': spin.[22]

I mentioned spin in Chapter 4. Electrons possess an intrinsic angular momentum characterized by a spin quantum number which is fixed at the value of ½. This means they can 'point' in two different directions in a magnetic field, which we have come to refer to as spin-up and spin-down. But, Penrose wanted to know, 'how does [the electron] know which way is "up" and which way is "down"?'[23] These are obviously directions we impose on the electron, derived from an assumed background spacetime.* Eliminating spacetime then means working only with the *total* spin angular momentum, deduced by combining all the spins of the objects in a collection. The total spin doesn't require the assumption of any kind of direction.

Penrose then imagined a universe constructed solely from objects carrying spin angular momentum. These objects are not

---

* Of course, the orientations of the electron spin are only revealed when the electron passes between the poles of a magnet, and so these poles impose a 'direction'. But this doesn't answer the question, as the magnet must still be orientated in spacetime.

assumed to be elementary particles: for now they are just 'objects'. The spin angular momentum is quantized, so the object can carry only fixed multiples of the fundamental unit of spin angular momentum, which happens to be an amount $\frac{1}{2}\hbar$, where $\hbar$ is Planck's constant h divided by 2π (about 1×10^{-34} joule-seconds). In other words, the object can carry angular momentum in units given by $n \times \frac{1}{2}\hbar$, where n = 0, 1, 2, 3, etc. The lines in Penrose's network are then labelled by the integer numbers n.[24]

The result is what Penrose referred to as a *spin network*, shown in Figure 19. The structure connects with physics by establishing some rules for how the objects can be combined, designed to ensure that total spin angular momentum is always conserved.

To see how this works, suppose we have an object carrying three units of spin angular momentum. This is represented as a line marked with the number '3'. The object doesn't move around (there is no spacetime, remember), but it can split or interact with other objects, so transferring and distributing spin angular momentum among a set of such objects. For example, the object carrying three units splits into an object carrying two units and

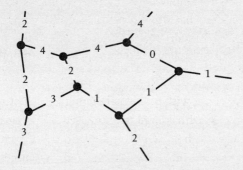

Figure 19. In an attempt to devise a description of space that is granular and relational, Penrose used the basic notion of spin angular momentum to develop spin networks, such as the example shown here.

another 3-unit, which then combines with a 1-unit to make a new 2-unit, and so on. This is all very abstract, but the point is that the only things of any significance in this structure are the topological *relationships* between the objects and their spin values.

Penrose showed that networks with sufficiently large total angular momentum serve to establish *directions* against which the relative orientation of other large networks can be measured. Imagine a spin network with a large value of the total spin angular momentum (N, say). Now transfer a single 1-unit to another large network (M). There is a certain probability that the 1-unit will add (to give $M + 1$), and a certain probability that it will subtract (to give $M - 1$). Penrose demonstrated that the relationship between these probabilities defines the *angle* between the two networks. What's more, he was able to show that all such angles are entirely consistent with angles in three-dimensional Euclidean space. Conventional space and geometry *emerges* as a set of relationships between large spin networks.

Penrose wasn't trying to devise a new formulation of quantum mechanics. He wrote:[25]

> I certainly don't want to suggest that the universe 'is' this picture or anything like that. But it is not unlikely that some essential features of the model that I am describing could still have relevance in a more complete theory applicable to more realistic situations.

Penrose had set out with no formal theoretical structure in mind and had come up with spin networks based simply on the notion that space *ought* to be discrete, and relational. In fact, Penrose's conception was not much more than an idea: spin networks did not constitute a research programme (Penrose was more involved with the development of another approach to quantum gravity, called twistor theory).

On discovering the solutions based on intersecting loops in Santa Barbara, Jacobson recalls his sense of excitement at the possibility of connecting these with Penrose's spin networks. Penrose had been present at the workshop in 1986, and Jacobson and Smolin had sought him out for a short tutorial on the subject. Penrose told them that he had prepared some handwritten notes and arranged to send them a photocopy. Jacobson had been thrilled and spent a lot of time learning how the scheme worked.

Smolin had certainly been aware of spin networks since his Ph.D. studies at Harvard. 'It is a mystery to me why it took so long before I realized that they were a basis for LQG', he admits.[26] But calculations involving spin networks appeared rather daunting. They were very intricate, and it was easy to make mistakes: a simple sign error in a long sum could be catastrophic. He nevertheless plucked up courage, and during a visit to Oxford in 1994 he sat down again with Penrose and learned how to do these calculations directly from their inventor.

It was an extraordinary achievement. Penrose had developed spin networks because the notion of a continuous spacetime didn't feel right. Quantizing Ashtekar's spin-connection formulation of general relativity had led first to the notion of 'loops' and 'knots' of gravitational force, then to a space constructed by weaving the loops together. Now Smolin and Rovelli had figured out how to apply the volume and area operators, and had shown that according to LQG space *is* discrete, and relational, with Penrose's spin networks staring right back at them.

Apply the operator for volume and you get discrete 'grains' or quanta of space *at the nodes* of a spin network. Apply the operator for area and you get discrete quanta of area *on the lines* of the network where adjacent grains touch (see Figure 20). The grains are incredibly small: about 10^{65} quanta of volume will fit inside a single proton. Moving from grain to adjacent grain in a closed

(a)   (b)

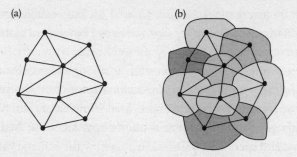

Figure 20. Smolin and Rovelli adapted Penrose's spin networks, (a) replacing the nodes by quanta of the volume of space and (b) the links with quanta of area.

circuit describes the 'loop' of the original form of LQG, and the sum of the values along the lines forming the loop gives a measure of the strength of the gravitational field and hence the extent of the *curvature* of space.

The journey had been at times confused, frustrating, and rather desperate. Although it was far from over, here was a rare moment of joy that Smolin and Rovelli could truly savour. They published two key papers, one on the discreteness of area and volume and the second on the connection between quantum gravity and spin networks.[27] These papers rank among their proudest scientific achievements and represent two of the most oft-cited papers on the subject of LQG.

It had been a productive summer.

But there was much still to be done. Over the next few years Smolin and Rovelli continued their collaboration with Ashtekar and joined forces with other theorists, including Pullin, Jerzy Lewandowski, Rodolfo Gambini (who, with Anthony Trias, had actually discovered the same loop methods quite independently and had sought to apply them to QCD), Renate Loll, Laurent Friedel, and Thomas Thiemann, among others. A long list of postgraduate students, postdoctoral associates, and visiting scholars would pass through the Center for Gravitational Physics and Geometry and go on to become leaders in the field.

Ashtekar and Lewandowski worked to establish a more rigorous and robust mathematical framework for the area and volume operators, deriving Smolin and Rovelli's earlier results, extending the volume spectrum, and developing some important new techniques.

Pullin contributed a critically important insight. He had earlier suggested that applying the volume operator to non-intersecting loops with no nodes would produce a zero-volume result.* Every node in the network is characterized by a certain number of quanta of volume. Adding up all the quanta of volume at all the nodes in a vast network results in a volume of space of the kind we experience. The equations demonstrated clearly that it is the *intersection* of the loops which is key. Without the nodes, there is no volume, and no space.

But there is a further subtlety. In Smolin's worst moment in science, Renate Loll stood at a conference in Warsaw in the summer of 1995 to declare that Smolin and Rovelli's calculation of the value of the smallest possible volume was wrong. The 'trivalent' network they had used—in which a single node is joined to three links—in fact has zero volume. 'The clarity regarding volume was slow to come about,' Rovelli admitted, 'and there was still some confusion because we got it wrong at first.'[28] After some 24 hours of frantic checking, Smolin found that he had made a single sign error in the calculation. Loll was right. The smallest possible volume is instead associated with a 'tetravalent' network in which a node is joined to four links.

---

* And the physical unreality of the zero-volume exact solutions that Smolin and Jacobson had discovered in Santa Barbara was now apparent.

But from this mistake the theorists learned a much simpler and more reliable method of performing the calculation than Penrose had taught them.

Ten years had elapsed since Ashtekar had used his 'new variables' to reformulate and greatly simplify the ADM Hamiltonian form of general relativity. Much water had passed beneath the bridge, and the theorists now had a physical interpretation for the solutions to their equations based on the volume and area of space as the observables of the theory.

There remained the rather vexing question of what to do about time.

10

IS THERE REALLY NO TIME LIKE THE PRESENT?

Carlo Rovelli and Lee Smolin published a paper summarizing their work on the discreteness of area and volume in the journal *Nuclear Physics B*, in May 1995. In it they declared: 'If one measured the volume of a physical region or the area of a physical surface with Planck scale accuracy, one would find that any measurement's result falls into the discrete spectra given here.'[1]

High-energy particle collisions recorded at the world's most powerful collider at CERN in Geneva are used to probe matter on distance scales on the order of an attometre (10^{-18} metres), and so measurements with Planck-scale accuracy (10^{-35} metres) are unlikely to become possible anytime soon. But there could be little doubting the nature of the theorists' achievement. General relativity and quantum theory had stubbornly refused to come together for at least 80 years. Now, through the efforts of a number of dedicated theorists, a resolution appeared to be at hand.

Remember, this was a theory that drew all the lessons about the nature of space and time from general relativity into a structure that closely resembled a quantum field theory. It produced a set of equations that could be solved without the need for messy and tedious renormalization procedures, and whose solutions could be shown to be completely independent of any choice of coordinate system—it was genuinely independent of any kind of spacetime background. What's more, this was a theory founded directly on accepted, empirically tried-and-trusted structures—there had been no need to second-guess physical reality any further by introducing supersymmetries, or hidden dimensions.

But any jubilation that the theorists might have experienced was short lived. Because, just a few months before Rovelli and Smolin's paper was published, some statements delivered to a conference by Edward Witten sparked a second superstring revolution.

Interest in the superstring programme had surged following the first revolution in 1984, rising to a peak in 1987 measured in terms of the number of publications and the number of citations of the original paper by Green and Schwarz.[2] But interest had begun to wane as the problems multiplied.

The most damaging problems related to the loss of uniqueness of the theory, in terms of both the number of superstring theory variants and the proliferation of possible Calabi–Yau spaces in which to hide the extra spatial dimensions, with no means to choose between them. The bandwagon had rolled very favourably for some, but job offers at prestigious institutions were beginning to dry up. By late 1994, several of the theorists who had been drawn to work on superstrings had become discouraged and had chosen to leave the field.

Some diehards remained. For these theorists, it seemed that the superstring programme was still too full of promise, the structure so beautiful that it surely had to have some role in the description of nature. Also, perhaps, there was already too much at stake: the programme was in some sense too big to fail.

Superstring theorists had managed to convince themselves that this was the *only* road to a theory of everything, that it was 'the only game in town'.[3] Many appeared to be blissfully unaware of other potential roads, if not to a theory of everything, then at least to a quantum theory of gravity.

There had been some rumblings from a few superstring theorists that, based on some calculations using supergravity, the right number of spacetime dimensions is actually eleven, not ten. It all seemed to be getting a bit out of hand. Then, at a string theory conference at the University of Southern California in March 1995, Witten stood to deliver a bold conjecture. Perhaps the five different ten-dimensional superstring theories and supergravity are actually instances or approximations of a single, overarching eleven-dimensional structure. He called it M-theory. He was not specific on the meaning or significance of 'M'.

But the superstring theories were formulated in ten, not eleven dimensions. Accommodating yet another dimension was possible only if the strings themselves were reinterpreted as higher dimensional objects, called *membranes*, or 'branes' (offering one possible interpretation of M, for membrane). These quickly took over from strings as the primary focus of investigation, leading British string theorist Michael Duff to refer to M-theory as 'the theory formerly known as strings'.[4]

This was a *conjecture*, not a theory. Witten demonstrated the equivalence of a ten-dimensional superstring theory and eleven-dimensional supergravity but he could not formulate M-theory; he could only speculate that the theory must exist. This is a simple fact that, in my experience, many readers of popular presentations of string theory somehow tend to miss. *M-theory is not a theory*. To this day, nobody knows what M-theory looks like, although many theorists have tinkered with structures that they believe it could or should possess. M-theory is really the *presumption* that there must exist a unique, eleven-dimensional superstring theory.

But this was, nevertheless, more than enough to set the superstring bandwagon rolling even faster. Superstrings were 'hot' once more, and interest surged. Institutions scrambled to recruit string theorists to their faculties.[5] Within just a few years, the theory had become the primary structure for exploring the 'big questions' of physics, eloquently explained in justifiably best-selling popular books, such as Brian Greene's *The Elegant Universe*, first published in 1999.

This was never necessarily about making a choice between the different approaches. Sound, scientific logic suggests that *all* roads should be followed to see where they lead, as indeed they were. But in the time since the second revolution the sheer amount of *effort* applied to these approaches was to become extraordinarily biased in favour of string or M-theory.

Developing a scientific theory is much like placing a bet. When faced with a choice between two actions, a rational person will choose the action which maximizes some expected utility. But, given that the world is a complex and sometimes unpredictable place, how do any of us know which action will achieve this? Should I keep my money in my bank account or invest in the stock market? I could try to determine a *probability* for each action to deliver the desired outcome and so choose the action with the highest probability. I don't necessarily need to calculate these probabilities: I might just look at bank interest rates and study the stock market and try to form an objective view, or I might run with a largely subjective opinion about these different choices, taking into account my appetite for risk.

Now, in any given year there's only ever a finite amount of money to go around to pay for academic positions, postdoctoral assistantships, student grants, travel bursaries, and computer time. In the USA, funding for the string theory programme was derived largely from the National Science Foundation's (NSF) high-energy particle physics budget and the Department of Energy. Funding for LQG was sourced elsewhere, from the NSF's gravitational physics division, so the two programmes were rarely in direct competition.[6] However, the gravitational physics division was also supporting LIGO, and difficult decisions on the allocation of funds will always come down to the relative merits of different programmes, no matter which budget they sit in.

If we *do* make this about a direct choice between superstrings and LQG, how might we judge the probabilities for these to deliver a quantum theory of gravity that might eventually become broadly accepted by the theoretical physics community as *the* answer?

I must confess that my personal judgement is guided by the nature (and the number) of the 'big ideas' that lie at the heart of each of these approaches. After all, this is where all the risk resides. The superstring programme is based on a confounding of not one but two big ideas—strings and supersymmetry—neither of which is grounded in any empirical evidence. Both ideas are derived from *speculations* about how nature *might* be, in response to problems with currently accepted structures, such as the use of point-particles and the hierarchy problem.

This is arguably not much different in status from the speculations of those ancient Greek philosophers who argued that nature *might* consist of atoms and void. Admittedly, the Greeks didn't have access to the elaborate mathematical frameworks available to modern string theorists, but in truth no amount of mathematics can turn metaphysical speculation into empirical fact. Only experimental science can do that.

To this we must add the problems of discovering precisely how to roll up all the hidden spatial dimensions that superstring theory requires. Glashow writes: 'Why, you may ask, do the stringers insist that space is nine-dimensional? It is not a consequence of elegant arguments nor compelling philosophy—it is simply that string theory doesn't make sense in any other kind of space.'[7] When we then add yet another big idea—M-theory—we begin to sense that what we're really building here is a rather elaborate house of cards.

In contrast, the only big idea behind LQG involves a mathematical technique for reformulating and greatly simplifying general relativity so that it looks much like a quantum field theory. All the rest—loops of force, knots, links and kinks, spin networks, and the quanta of area and volume—are derived through the application of techniques that are already broadly familiar in the quantum field theories of the standard model of particle physics. LQG had been built from what we already *know* about nature, and can prove, from general relativity and quantum theory.

Yes, these structures must be twisted and tortured so that they can be brought together, and we lose some confidence in their integrity in the process. Some argue that string theory is the tighter, more mathematically rigorous and consistent, better-defined structure. But a good deal of this consistency appears to have been imported through the assumption of supersymmetry, and with each survey published by the ATLAS or CMS detector collaborations at CERN's Large Hadron Collider, the scientific case for supersymmetry weakens some more.

I would argue that it is surely better to start with nature as we understand it, not with how we think (or wish) nature might be, even if this means sacrificing some mathematical rigour and consistency. Even a passing acquaintance with science history will reveal that a lack of mathematical consistency has never stood in the way of a genuine discovery. The road from relativity requires a modest leap of faith and, to all the world, looks like the 'safe bet'. This approach is much more consistent with the strategy that was used to build the highly successful structure of quantum field theories that make up the current standard model of particle physics.

Of course, LQG also has many shortcomings. In 1995, superstring theory still promised a theory of everything, whereas the road from relativity promised *only* a quantum theory of gravity.* As Conlon writes, 'string theory has proven to be so much more than just quantum gravity—and by doing so it has become attractive to large numbers of scientists.'[8] Particle theorists brought up on a diet of quantum field theory rightly wanted to know what had become of the *graviton*. Familiar and comfortable working with a background spacetime (or simply ignoring that this might be a problem), they were also unnerved by the loss of the time dimension.

And, sadly, neither approach was in a position to provide the kind of prediction and empirical test that might allow experimental physics an opportunity to reveal nature's own preferences. There was simply no practical means to choose between them, and as the physics these theories describe happens at unimaginably small dimensions, or incredibly high energies, any kind of contact with experiment seemed unlikely for quite some time to come.

But if we are to understand what really happened to theoretical physics after 1995 we must inform this somewhat pragmatic rationale with what we know about human psychology. In his book *The Trouble with Physics*, first published in 2006, Smolin

* Smolin never really understood this argument, as LQG readily accommodates the quantum field structures demanded by the standard model of particle physics. It can also incorporate Higgs fields and even supersymmetry, if required.[9] For his part, Rovelli retorts: 'we are not near the end of physics, we better not dream of a final theory of everything, and we better solve one problem at [a] time'.[10]

identified a series of traits that he argued are characteristic of the string theory community.[11] These included the following:

- Tremendous self-confidence (I would add: 'at times bordering on arrogance');*
- An immensely strong group consensus and sense of identification ('group-think'), with sharply defined boundaries distinguishing members from 'others' outside the group;
- A complete disregard and disinterest in the opinions of anyone outside the group; and
- A tendency to interpret evidence optimistically (I would go further, and declare this for what it is: *confirmation bias*—a tendency to focus attention only on arguments that support or confirm the established preconceptions of the group).

Smolin hints at it, but of course these are traits that we would tend to associate with particularly ardent members of a political party, Church, or a cult.† Already in 1988, Sheldon Glashow was writing: 'According to this new religion' and 'String thoughts may be more appropriate to departments of mathematics or even to schools of divinity. How many angels can dance on the head of a pin? How many dimensions are there in a [Calabi–Yau space] thirty powers of ten smaller than a pinhead? Has medieval theology been resurrected?'[12]

Yes, we can base our judgements about which horse to back in the race to develop a quantum theory of gravity entirely on

---

* One of my favourite quotes from Canadian philosopher and 1960s media theorist Marshall McLuhan reads: 'I may be wrong, but I'm never in doubt.' For more 'McLuhanisms', see: https://marshallmcluhan.com/mcluhanisms/.

† I should point out that these are observations, not criticisms, per se. Many human beings in all walks of life have a tendency to behave like this, and these traits may indeed be ascribed to some members of the LQG community.

rational criteria, but it would be extremely foolish to overlook the fact that science is a human endeavour. Such judgements also inevitably depend on rather more irrational considerations, derived from the perceptions, culture, and value sets of the community's leaders, and the way in which they exercise their power and influence. Those with recognized authority tend to set the standards by which the value of research efforts are judged and rewarded. They can greatly influence funding decisions and help to compete for the best minds at the world's most prestigious academic institutions.

Rovelli recalls his experience at a high-profile annual string theory conference held at CERN in 2008. His invitation to attend had caused some consternation among the string theory faithful, despite the general practice of routinely inviting string theorists to attend conferences on LQG. While the young string theorists were interested in what Rovelli had to say, in his summing up David Gross was very negative: 'String theory absorbs everything happening in theoretical physics. Except LQG. But we are not interested in absorbing LQG.'[13]

There's a saying I've come across in the business world: 'Whatever interests my boss fascinates me.' If the string community is a Church, then there can be little doubting that Witten is its Pope. In *Why String Theory?*, Conlon acknowledges Witten's influence: 'What he worked on was automatically fashionable— if he thought something was important, it was. There were many other smart people in the subject, but there was only one Big Ed.'[14] Smolin recounts some almost comical instances of string theorists suddenly becoming fascinated by a seemingly unrelated development, because they'd seen Witten in the library, reading about it. Smolin began to be infuriated by the phrase 'Well, what does Ed think?', commonly heard in conversation after a conference presentation.[15]

With the demand for empirical evidence pushed firmly into the background, it becomes perfectly acceptable to ignore the demand completely, to lose all respect for it, and, arguably, to abandon what passes for the scientific method. The real danger is that, when cast adrift in this way yet with lots of very smart theorists arguing in support, research programmes can become self-perpetuating, even when they appear to have failed to deliver on whatever promise they held.

Ashtekar had hoped that the reformulation of general relativity based on his new variables would remove the 'wedge' between those travelling the road from relativity and those on the road that had led to string theory. Smolin saw hope for a unification of superstring theory and LQG, and developed a programme to attempt this (this hope was the basis of his book *Three Roads to Quantum Gravity*). He spent a decade from 1995 to 2005 working on a background-independent version of M-theory or, equivalently, an M-theory extension of LQG. But the second superstring revolution and what followed drove the wedge even deeper, and Smolin's efforts were largely ignored on both sides as eyes averted and ears became deaf.

The result was nothing less than a schism.

Penrose had intended his spin networks to serve as a framework from which spacetime would emerge naturally. But in LQG the nodes of the spin network are where we find the grains or quanta of the volume of *space* only, and the links between the nodes are characterized by the quantum numbers for area where the grains touch. This is very much a static picture. Somehow, the theorists had to figure out a way to introduce some kind of dynamical change. It became known as the *problem of frozen time*.

For sure, philosophers have pondered on the reality of time for centuries. Through the special and general theories of relativity, Einstein had demonstrated that time is relative, not absolute.

He had no hesitation in declaring: 'For those of us who believe in physics, the distinction between past, present and future is only a stubbornly persistent illusion.'[16]

This is all very well, but despite what some philosophers, physicists, and a few relatively obscure mathematical equations might say, there's no denying our personal experience of the passage of time (and we'll be returning to this subject in Chapter 16). There's also no denying the importance that time plays throughout science, and in most of physics. General relativity is, after all, a theory of spacetime interpreted as the gravitational field.

In a theory from which time had disappeared, how were the theorists supposed to put it back?

The trick was, once again, to set aside prejudices born from familiar experience and turn the problem on its head. We're conditioned by our experience always to think of space and time as a 'backdrop' against which events in our physical world play out. Imagining a world without space or time then requires a very demanding form of mental gymnastics.

In our world of experience, we see this happen, then that. Provided we can discern a rational, physical connection between the two events, we readily conclude that this *caused* that. When we see the same kind of thing happening again and again, we further conclude that this always *causes* that.

Quantum theory disturbs this simple logic somewhat, but don't be fooled. Einstein's concerns about a dice-playing deity notwithstanding, quantum theory does not completely undermine the relationship between cause and effect. In our classical world of experience, we learn that this causes that, with certainty. In the quantum world, we have to learn to be more circumspect. We learn that this causes that with a certain quantum probability, one that we can calculate from the wavefunction which describes whatever 'this' is. This may also cause something else to happen with a different quantum probability, such that the probabilities sum to 100 per cent. We don't know exactly what will happen—that or something else—until we look (or make a measurement), at which point the wavefunction collapses and we get one or the other. But, make no mistake, one or the other *will* happen. There are still some things we can be certain about.

In a world without time, how should we then think about cause and effect? We might conclude that without time, there can surely be no causation, because nothing ever happens. But cause and effect is nothing more than a *relationship*, and what is LQG but a theory built entirely on relationships?

A spin network gives a static picture. But let's now stack another, slightly different, network on 'top' of this, and trace the relationships between the nodes and links of one to the other. On paper, the two-dimensional graph of nodes and links now becomes a three-dimensional object in which the nodes become edges and the links become faces. The edges 'carry' a certain volume of space, and the faces 'carry' a certain area. We add another network on top of this, then another.

For example, suppose we have a spin network which consists of three nodes (three quanta of volume) and six links (six quanta of area). We overlay another network with a single node and three links. 'Mapping' the evolution of the nodes and links then describes a *transition* between different networks or different quantum states of space (see Figure 21).

The result is what the theorists learned to call a *spinfoam*. Why 'foam'? This is not a smooth and continuous process, but rather characteristically quantum in nature. Space evolves in fits and starts, 'jumping' from one configuration to another. Based on little more than intuition, John Wheeler had imagined what space-time might be like at the Planck scale: 'Paradoxically, when I later became interested in gravitation and general relativity, I found

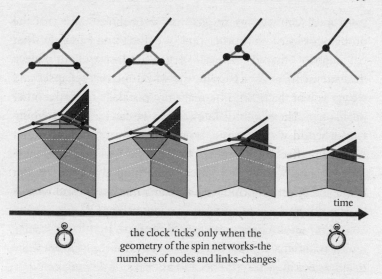

time

the clock 'ticks' only when the
geometry of the spin networks-the
numbers of nodes and links-changes

Figure 21. In a spinfoam the nodes of the spin network become lines and the links become planes. In this example, a spin network evolves from 3 quanta of volume (3 nodes) and 6 quanta of area (links) to 1 quantum of volume and 3 quanta of area. This evolution is not continuous, however. What matters are only the *numbers* of nodes and links, so the clock 'ticks' only when these change.

myself forced to invent the idea of "quantum foam", made up not merely of particles popping into and out of existence without limit, but of spacetime itself churned into a lather of distorted geometry.'[17]

Now, the two-dimensional graph of the spin network actually represents three-dimensional space. So by mapping the nodes and links from one network to the next—in other words by establishing evolving causal relationships between slices of frozen time—we have, in fact, introduced a fourth dimension. Let's call it time. Each spinfoam then traces a 'history' of local events involving transitions between the quantum states of space. How these are drawn is not so important. What is important is the *geometry* of the network in terms of the numbers (and values) of the nodes and links. The transition shown in Figure 21 does not happen continuously—only in the first and last graphs is the geometry different and the clock 'ticks' only when the geometry changes. Such transitions happen pretty quickly, on the order of the Planck time, which is the time it takes for light to travel the Planck length, about 5×10^{-44} seconds.

The details of the spinfoam model were hammered out in a series of papers published from 1995 through to 2000. The history is quite complex. Aside from Rovelli and Smolin, a number of other theorists were involved, including John W. Barrett and Louis Crane, John Baez (who coined the term 'spinfoam' in tribute to Wheeler's vision), Laurent Freidel, Junichi Iwasaki, Kirill Krasnov, Fotini Markopoulou, and Michael Reisenberger.

This might make some sense, but we could be forgiven for thinking this is no more meaningful than playing games with a few graphs. It all looks rather arbitrary. How are we meant to determine from this precisely how the quantum states of space really do evolve? The answer was to reach for an older—and very familiar—approach to quantum mechanics that Richard Feynman had developed in 1948.

The basis of Feynman's approach can be found in some of the simplest observations of classical physics. The pencil-thin beam of light from a laser travels in a straight line. Why? Because straight lines represent the least amount of time required for light to travel from its source to its destination, first enunciated as a principle by Pierre de Fermat (of 'last theorem' fame) in 1657. But how is the light supposed to 'know' in advance what the path of least time is?

Feynman's answer was that light does not need to know the path of least time in advance, because it takes *all paths* from its source to its destination.

Not sure? Then let's imagine inserting a screen between a point source of light and a photographic plate. If we drill a hole in the screen then the amplitude or intensity of the light that reaches the photographic plate is simply related to the amplitude of the light passing through the hole. If we now drill a second hole, we obviously need to sum the amplitudes of the light passing through both holes. We can keep repeating this for three, four, five holes, each time summing the amplitudes for all the different possible paths through the screen.

By the time we have drilled an infinite number of holes in the screen, the screen is no longer there. We conclude that we must still sum the amplitudes of all the possible paths from the source to the photographic plate.

Feynman elevated these relatively simple physical principles into an alternative formulation of non-relativistic quantum mechanics. He represented the passage of a quantum particle from one place to another as a sum over all the possible paths the particle could take or, alternatively, as the sum over all possible 'histories' of the particle's motion. The probability of finding the particle at a specific location is then determined from the amplitudes of all the various paths it can take to get there.

In the quantum world, 'taking a path' doesn't simply mean moving from point A to point B, as all manner of exotic things can happen on the way, involving the creation and annihilation of virtual particles, as I described in Chapter 4. Each of the various possible 'paths' or histories can be pictured as a Feynman diagram. The sum over histories is then equivalent to a sum over all possible Feynman diagrams.

Of course, Feynman diagrams are drawn against an assumed background spacetime. But the theorists discovered that much the same approach could now be applied to LQG. Each spin-foam represents a possible history—a sequence of steps tracing the causal relationships between the nodes and links of the networks—or the quantum gravity equivalent of a Feynman diagram. Each spinfoam is associated with a probability. To know how the quantum states of space evolve it is necessary to compute the probabilities for all the different possible spinfoams—all the different possible histories—and then sum over these. Adding up all the different spinfoams creates a quantum superposition. We call the large-scale end result *spacetime*.

Now, there are many open problems with the spinfoam approach, and I'm sure you won't be surprised to learn that it's not as straightforward as I've described it here. Spinfoams allow us to conjure spacetime from evolving spin networks *in principle*, but please be under no illusions: this is still a very active area of research.

For theorists interested in the road from relativity, the Center for Gravitational Physics and Geometry that Ashtekar and Smolin had helped to establish at Penn State in 1993 had remained a fundamentally important island refuge in an ever-growing turbulent sea of string physics. It drew in many bright young theorists to work on what had become the less fashionable approach to a quantum theory of gravity.

One such theorist was Fotini Markopoulou. Her work on the causal evolution of spin networks had caught Smolin's eye in 1997. Raised in Athens, Greece, Markopoulou had graduated from Queen Mary College in London and had studied for her Ph.D. under Isham at Imperial College. Smolin invited her to join the group at Penn State for a few months as she completed her dissertation. She returned to Penn State the following year, recruited by Ashtekar, Smolin, and Pullin.

In a field already full of non-comformists (by definition), Markopoulou was something of an extremist. She had rejected string theory because of its 'strong machismo', and at Penn State she found herself in the right place, at just the right time: 'A bunch of different ideas were coming together; there was this sense that you might actually do something faster than the person in the next room, which is very unusual in quantum gravity,' she explained.[18] But, having helped to establish a technique which could bring the illusion of time back into the LQG picture, she rejected this basic premise, insisting that it is *time* that should be considered to be fundamental, with space merely an impression left by time's passing. Smolin was initially resistant to this idea, but the more he thought about it, the more he was inclined to agree, and the nature and status of time would become a major theme of his research in the coming years (to which we'll return in Chapter 16).

His work with Markopoulou and their shared vision of physics drew them close. They married in 1999, and together moved to Imperial College in London, returning a couple of years later.

In 2000, the superstring bandwagon was in full roll, attracting a lot of attention and a lot of funding. Smolin was warned by Kip Thorne that anything not related to the search for gravitational waves was likely to be squeezed out of the NSF's gravitational physics budget.[19] Things were starting to look bleak.

Smolin and Markopoulou were thus greatly (but pleasantly) surprised to be approached by Howard Burton, a young theorist who had only just completed his Ph.D. at the University of Waterloo in Ontario, Canada. Burton didn't want to talk to them about a post-doctoral position or a research grant. He wanted to know whether they were interested in helping to found a whole new *institute* for theoretical physics.

Casting around in search of gainful employment, Burton had brazenly written letters directly to a number of CEOs asking about job opportunities. His letter had come to the attention of Mike Lazaridis, the founder and co-CEO of Research In Motion, Inc., the manufacturer of the BlackBerry smartphone.

It so happened that Lazaridis did have a job opportunity. He was thinking about setting up a new institute for theoretical physics and wondered if Burton might be interested in researching his proposal.

It was not Burton's extraordinary story that piqued the interest of Smolin and Markopoulou (at their first meeting he didn't mention who the benefactor was). It was his suggestion that there might be considerable funds to support the new institute, and in further exchanges a figure of $100 million was mentioned.

As part of his research, Burton had talked to many theorists. As a Ph.D. student, he had been shocked to discover that in quantum gravity research 'the field was rife with dissention and sociological barriers. Superstring theorists, for example, did not interact in any meaningful way with people pursuing other approaches to the problem, and vice versa'.[20] Over dinner that evening, Smolin and Markopoulou expressed considerable interest in the new venture, and subsequently travelled in secrecy to Canada, where Lazaridis himself talked to them about joining what was to be called the Perimeter Institute for Theoretical Physics.

Smolin was in transition, having taken two years' unpaid leave from Penn State to work with Isham at Imperial College in London.* Both Smolin and Markopoulou agreed to join the new institute and research operations began in 2001 with a founding faculty of just three—Smolin, Markopoulou, and Canadian string theorist Robert Myers—and four post-doctoral assistants. By this time Smolin and Markopoulou had agreed an amicable separation.

---

* Smolin was funded by a private foundation that was seeking to establish a professorial chair for him at Imperial College.

There was clearly something about the approach of a new millennium that invited us to consider big, life-changing decisions.* The wind of change was in the air for Rovelli, too. Quite by chance, in 1994 he had found himself at dinner at the Isaac Newton Institute for Mathematical Sciences in Cambridge, England, sitting next to the celebrated French mathematician Alain Connes. It was an enjoyable meeting, fueled by several glasses of wine, during which they talked about a number of subjects in mathematics and physics. At one point, Connes confessed: 'I have an idea about how time emerges, but nobody has taken me seriously.'[21]

After some confusion, Rovelli realized that Connes' arguments were entirely consistent with the work he himself had been doing, and he rushed upstairs to fetch copies of his recent papers. They pored over them at the dinner table. They had argued from different directions, using different mathematical techniques, but Connes saw that Rovelli's was simply a special case of his own approach. They published a paper together later that year.

This was to prove a providential collaboration. Being associated with Connes opened a few doors to Rovelli in France. A few months after their joint paper had appeared, Rovelli received a phone call inviting him to join the Centre for Theoretical Physics in Luminy, in the ninth arrondissement of Marseille, perched on the Mediterranean coast.

He worked in Marseille as Director of Research at the centre in 1998 and 1999, and eventually left Pittsburgh in 2000 to take up a professorship there. His ten-year American adventure had allowed him to continue his career in science and had provided

* I'm taking the majority, populist approach in considering 31 December 1999 as the last day of the last millennium. This was an important time for me, too. I chose to leave corporate life with Shell to set up my own business. I had lots of reasons, but one of them was so that I could find more time to write about science.

him with new lifelong colleagues and some great opportunities. He had many regrets about leaving. But this time he was much less hesitant about the decision.

Ashtekar, Rovelli, and Smolin were once more scattered across the globe, though their commitment to collaborate was undiminished. They had come a long way together, and they could be immensely proud of what they had achieved. But there was much still to do. Some years later, Penrose would summarize his take on LQG this way:[22]

> I have little hesitation in saying that these developments are the most important in the canonical approach to quantum gravity since the subject itself was started roughly half a century ago, by Dirac and others. The loop states do appear to address at least some of the profound problems raised by general covariance. Moreover, these developments seem to have moved the discussion in a fascinating and perhaps not fully anticipated direction, where some gratifying elements of discreteness in spacetime structure begin to appear.

In his popular book *Three Roads to Quantum Gravity*, first published in 2000, Smolin was even more optimistic. He suggested that we would have 'the basic framework of the quantum theory of gravity by 2010, 2015 at the outside'.[23] He went on to predict that the theory of quantum gravity would be taught to high school students by the end of the twenty-first century.

Much had been achieved but as the millennium turned it was clear that there was still plenty to do. Perhaps most importantly, LQG was intended to be a theory of gravity, and gravity is manifested as a 'force' acting between material objects.

It was time to put some *substance* into the new theory.

PART III
ELABORATIONS

11

GRAVITONS, HOLOGRAPHIC PHYSICS, AND WHY THINGS FALL DOWN

What is the world made of? The answer is, perhaps, one of the few things that the relativists and string theorists can broadly agree on. If by 'world' we mean 'everything', then the answer is, unequivocally, *quantum fields*.

Of course, quantum field theory cannot explain literally everything. We don't routinely reach for the theory to explain the intricate details of human evolution, consciousness, or domestic politics, so we probably ought to qualify what we mean by 'everything'. What we really mean is that the fabric of our physical reality—space, time, matter, and energy—is, at its lowest level (and at least for now), reducible to a series of quantum fields and their interactions. The standard model of particle physics describes the properties and behaviour of elementary particles and the forces between them in terms of quantum fields *in* spacetime. Loop quantum gravity then insists that the properties and behaviour of spacetime itself is described by another, background-independent quantum field from which spacetime emerges.

To paraphrase the claims of the little old lady at the back of the room, as described either by philosophers William James or Bertrand Russell (take your pick): It's quantum fields *all the way down*.[1]

Whilst there was still a lot to do, the basic logical and mathematical framework of LQG was more or less in place by the mid–late 1990s. Many theorists had worked to confirm, resolve, refine, and extend the theory's mathematical foundations and its principal conclusions in terms of spin networks, quanta of area and volume, and approaches to emergent spacetime based on spinfoams. Notable among these was Thomas Thiemann, a young German theorist who had started out at the Institute for Theoretical Physics in Aachen, and had moved to Penn State to work with Ashtekar before going on to Harvard.

By 2000 Abhay Ashtekar, Carlo Rovelli, and Lee Smolin were separated physically, but in truth their intellectual separation had started some five years earlier. They continued to collaborate, but their research interests had begun to diverge. From about 1995, Smolin and Rovelli developed different ideas about how LQG could progress. Ashtekar and his collaborators, including Thiemann and Lewandowski, worked strenuously to give LQG a more precise mathematical prescription, and laid some fundamentally important structural foundations. Towards the end of the 1990s, Ashtekar's focus turned to aspects of general relativity relevant to cosmology and the physics of black holes.

Smolin was keen to bring LQG into the broad ambit of modern theoretical physics, to make it mainstream by linking it to other field theory structures, and especially the string theory programme. Despite their differences, Smolin sensed that all these approaches had more in common than many theorists appeared willing to admit, suggesting the promise of a deeper set of relationships between them. Finding these would require something of a grand vision and synthesis, and Smolin was heavily influenced in his thinking by discussions with Louis Crane.

Smolin set out his stall in what he refers to as his 'linking' paper, published in the *Journal of Mathematical Physics* in 1995, 'which I think of as my best solo contribution to LQG...I am very proud of this paper, which I think was ahead of its time, indeed, several of its results...have been recently rediscovered.'[2]

Rovelli believed that LQG was now a consistent, self-contained structure and he was keen to see the new theory through to some kind of sense of finality. This meant bringing in techniques borrowed from quantum field theory in the hope of uncovering some new physics that might provide an empirical test, one that could perhaps settle the matter of the nature of space and time, once and for all. In a series of introductory lectures to graduate students delivered in Zakopane, Poland, in 2011, he said that LQG:[3]

> must be judged on the basis of two criteria. The first is whether it provides a coherent scheme consistent with what we know about Nature, namely with quantum mechanics and, in an appropriate limit, with classical general relativity. The second is to predict new physics that agrees with future empirical observations. This is all we demand of a quantum theory of gravity.

These views are not incompatible, of course, but they demand different research priorities and different commitments. As the new century progressed, Smolin and Rovelli went separate ways. Rovelli would take occasional timeouts to make pronouncements on the interpretation of quantum mechanics (as we will see in Chapter 13), but his principal interest was in turning LQG into a broadly accepted foundational theory of physics. He wanted to finish what he'd helped to start.

As we learned in Chapter 6, the alternative covariant approach to quantum gravity *starts* with the assumption of a graviton field sitting on a flat background spacetime. The graviton field is then believed to be responsible for the spacetime curvature that we associate with gravity in general relativity.

But because a background spacetime is completely absent in LQG—by design—the theory doesn't make use of gravitons as building blocks and consequently these particles don't make themselves quite so obvious. Perhaps inevitably, particle theorists are naturally biased towards the idea of particles as the corollary to quantum fields. So, if LQG is really a quantum field theory then— the particle theorists insisted—there really ought to be quanta of this field that are particles. Where were they? We needed to know.

A related question concerns the consistency of LQG with general relativity and, ultimately, with the classical Newtonian understanding of gravitation, with its inverse-square law and apparent action at a distance. Einstein had held great store by his principle of consistency. It was important that he should be able to recover Newton's inverse-square law as a limiting case or approximation of general relativity in situations in which we can assume a flat spacetime and speeds substantially below light speed.

So, would a limiting, low-energy case or approximation of LQG recover general relativity? This is, after all, a theory of *gravity*. Could we then trace a straight line in the logic that begins with quanta of area and volume and transitions between networks in spinfoam all the way to our own, distinctly human, classical intuition for the reason why things fall down? This certainly should be possible, in principle.

As I explained in Chapter 4, the understanding derived from the standard model of particle physics is that forces between matter particles such as quarks and electrons are 'carried' by force particles (look back at Figure 10 in Chapter 4). The photon carries the electromagnetic force between positive and negative electrically charged particles. Electromagnetism is a long-range force, transmitted by massless force carriers which are the quanta of the electromagnetic field.

The W and Z bosons carry the weak force between different flavours of quarks and leptons, such as electrons and neutrinos. The weak force is short range, at work inside the nucleus, transmitted by heavy force carriers which we can think of as 'heavy photons'. These are the quanta of the 'weak field' (more generally called the 'weak interaction'). Finally, the gluons carry the strong colour force between differently coloured quarks, such as red and blue. Although also short range, as we have seen, the colour force works very differently and the gluons are massless, like the photon. They are the quanta of the colour field.

You get the basic idea.

Against the background of this kind of understanding, the notion that the 'force' of gravity must also be transmitted by force carriers, the quanta of the gravitational field, appears pretty irresistible. If, as many theorists insist, gravity is also a quantum phenomenon, then surely gravitons must exist. We know that gravity is a long-range force, so this implies a massless force-carrier. The nature of gravity further suggests that the graviton should be a 'tensor boson', with a spin of 2.*

In a Feynman diagram, the force carrier passes as a virtual particle (drawn as a wavy line) between matter particles—such as quarks and leptons—that are susceptible to the force (look back at Figure 8 in Chapter 4). A logical extension of this picture to gravity would have the gravitational force, carried by a

---

* The Higgs field is a so-called scalar field (it doesn't 'point' in any particular direction) and the Higgs boson therefore has a spin of 0. Photons and the W and Z particles are 'vector bosons' with spin 1.

virtual graviton (drawn as a wavy line), passing between matter particles.

Although this is all very persuasive, this is *not* the conclusion of LQG. The fundamental quanta that emerge from this formulation of the quantum theory of gravity are not gravitons. As we have seen, they are instead quanta of the *area* and *volume of space*.

These very different views are difficult to reconcile. Particle theorists see the graviton as essential in any quantum theory that purports to describe the force of gravity. In contrast, general relativity does not interpret gravity as a force at all, but rather as a secondary consequence of the distortion of spacetime geometry (the gravitational field) by the presence of mass-energy. Of course, this doesn't mean that *changes* in the gravitational field have no physical effect—they obviously do, as Rovelli explains:[4]

> If the Sun is suddenly swept away, the disturbance in the local spacetime geometry travels for eight minutes (at the speed of light) until it reaches the Earth, and we feel the change. And in the meanwhile a disturbance in the electromagnetic field travels for the same eight minutes until it reaches the Earth, and we see that the Sun has disappeared. The second effect is 'carried' by the electromagnetic field, the first is 'carried' by spacetime.

It probably shouldn't come as a great surprise to find that a quantum theory of gravity deduced by following the road from relativity doesn't feature the graviton as a fundamental excitation or vibration of the quantum gravitational field. In a canonical quantum gravity, there is basically no force to be carried, as such. But this doesn't mean to say that there are *no* gravitons in LQG.

One of the principal lessons from the birth and early childhood of quantum mechanics in the 1920s is that nature exhibits a basic duality. We learned that certain phenomena described exclusively in terms of waves (such as light) can also be described as particles. Likewise, certain phenomena described exclusively in terms of particles (such as electrons) can also be described as waves. As these physical descriptions matured, we replaced 'waves' with 'fields' and started to think about 'particles' as fundamental excitations or vibrations of the fields rather than self-contained bits of material substance.

But this essential duality in nature is not confined to elementary particles and their associated quantum fields. *Any* phenomenon at the atomic level that can be described as waves that carry energy in some form can, in principle, also be interpreted in terms of associated particles. Think of it this way. A wave carrying an energy E will be associated with a certain mass m given by $m = E/c^2$. The 'mass of a wave' doesn't make much physical sense: instead we think of mass as a particle-like property.*

Now, we don't tend to associate macroscopic wave phenomena—such as waves in the ocean—with particles. But when we're working at the microscopic level of atoms and subatomic particles then the particle-like aspects of wave phenomena (and the wave-like aspects of particle phenomena) become more and more important, for the simple reason that at this level Planck's constant h, though small, can no longer be ignored.†

Some solids and liquids are formed from a regular three-dimensional array or lattice of atoms. These atoms are not

* For particles with zero rest mass, such as the photon, we must reach for the fully relativistic expression for the energy, E, from which we can deduce that $p = E/c$, where p is the linear momentum (also a particle-like property).[5]

† The de Broglie relation connects the wave property of wavelength (λ) and the particle property of momentum (p) according to $\lambda = h/p$, where h is Planck's constant. For macroscopic objects such as tennis balls, for any measureable value of momentum the wavelength will always be much, much shorter than the shortest X-rays or γ-rays. However, for microscopic objects such as electrons, it becomes possible to discern *both* particle-like and wave-like behaviours depending on the experimental setup, as we saw in Chapter 3.

completely frozen in place in the lattice, but have some limited 'wriggle room'. Small movements of one atom can be communicated to adjacent atoms, passing the movement along the lattice. The result is a collective vibration which we can describe as a wave. Such vibrations may have a range of frequencies, but these can be decomposed into a set of fundamental frequencies that represent the standing waves (called 'normal modes') characteristic of the substance.

It makes sense to describe these vibrations in terms of waves passing along the lattice. But these waves are derived from microscopic phenomena in which Planck's constant h can't be ignored, and in quantum mechanics we can choose to describe these waves also as particles. In 1932 the Soviet physicist Igor Tamm used the name *phonon* to describe the particle equivalent of these lattice vibrations. He chose this name because some of the frequencies of vibration lie in the audio range (they are sound waves). Phonons are sometimes referred to as *quasiparticles*. They are not elementary particles, and they don't carry a force in the way that photons, W and Z particles, and gluons do.

Let's get back to gravity. The detection of gravitational waves by LIGO in 2015 was a fantastic achievement, but many physicists familiar with general relativity and the properties of black holes figured that it was really only a matter of detector sensitivity—and therefore time—before gravitational waves would be found.

In quantum mechanics we can choose to describe waves also as particles, and of course the particles associated with gravitational waves are gravitons.

How does this help? LQG predicts the quantization of space into discrete volumes, which sit on the nodes of the spin networks. In a spinfoam the spin networks transition (or 'jump') from one to another and the spacetime that emerges by summing over the spinfoams has a 'natural' lattice structure, determined by the discrete quantum nature of space.

This is very different from the imaginary lattice we impose on the physics just to make it easier to perform difficult calculations in QCD. For one thing, the natural, emergent lattice is dynamic, whereas the imaginary lattice of QCD is static. So, we can think of gravitational waves as collective vibrations transmitted along the spacetime lattice. Seen from this perspective, gravitons are quasiparticles associated with the collective vibrations of spacetime, much like phonons are quasiparticles associated with collective vibrations in certain solids or liquids.

We can now close the circle. If the Sun suddenly disappears, this disturbance in our local gravitational field is transmitted by gravitational waves which take eight minutes to reach the Earth. We can associate these waves with massless gravitons, travelling at the speed of light.

As we've seen, superstring theory predicts a 'closed loop' string with properties characteristic of the graviton. Ashtekar, Rovelli, and Smolin had deduced already in 1991 that LQG predicts gravitons as quasiparticles associated with gravitational waves. 'We knew pretty early that LQG had gravitons', says Smolin.[6] On this basis, the odds look pretty good that gravitons must exist. Furthermore, they really ought to be ubiquitous given that we're constantly surrounded by a spacetime that is filled with gravitating objects. So how come we've never seen them?

We probably shouldn't get too carried away. In his review of Brian Greene's *The Fabric of the Cosmos*, published in *The New York Review of Books* in May 2004, English theorist Freeman Dyson* used the opportunity to address a rather interesting question: Is the graviton detectable, *in principle*? He wrote:[7]

---

* Dyson rose to fame in the late 1940s by demonstrating the equivalence of different versions of QED developed by Feynman, Schwinger, and Japanese theorist Sin-Itiro Tomonaga. His work paved the way for the establishment of QED as the accepted quantum field description of electromagnetism.

Because of the extreme weakness of the gravitational interaction, any putative detector of gravitons has to be extravagantly massive. If the detector has normal density, most of it is too far from the source of gravitons to be effective, and if it is compressed to a high density around the source it collapses into a black hole. There seems to be a conspiracy of nature to prevent the detector from working.

This wasn't the first time this problem had been raised. Dyson himself had mentioned it earlier and Smolin had published a couple of papers in the early 1980s in which he had concluded that gravitons would be undetectable. A couple of years later Tony Rothman and Steven Boughn at Princeton University explored potential answers to Dyson's question in some detail. They concluded that there is nothing in the hypothetical physics of gravitons that would necessarily prevent their detection. But any experiment would require a detector the size of Jupiter, operating at a quite unrealistic 100 per cent efficiency, placed in orbit around a white dwarf or neutron star.

If this isn't impractical enough, to prevent the detector from being overwhelmed by 'noise' caused by the detection of neutrinos, the apparatus would require a shield with a thickness measured in light-years. The shield would quickly collapse to a black hole. We can speculate that any proposal to build such an experiment is unlikely to attract much sympathy from Earth-bound funding agencies (or the taxpayers that support them).

According to Rothman and Boughn the answer then is a pretty conclusive 'no': 'we do feel entitled to predict that no one will ever detect [an individual graviton] in our universe'.[8]

We logically infer that when we detect gravitational waves we're detecting gravitons in large numbers, and we may yet be able to deduce some properties of gravitons. But the practical impossibility of detecting individual gravitons means that, for now, the notional particle of the gravitational field remains firmly hypothetical.

This answers the graviton question but it immediately begs another. How then should we model the motions and interactions of *particles* in LQG? We can address this question in the context of the graviton but, of course, it applies equally to any standard model particle—such as an electron—whose physics we might want to describe using a structure that does not presuppose a spacetime background.

To a certain extent, we could simply suggest that the quantum nature of space and time predicted by LQG is manifest only at the Planck scale, and the physics described by the standard model works on scales that are some 17 orders of magnitude larger than this. As far as the standard model is concerned, surely spacetime can be *assumed* to form a continuous background on which this physics happens?

But this really isn't good enough. If we're going to invest any belief in what LQG is telling us about the structure of space and time at the Planck scale, then the theorists need to *show* how more familiar particle physics might emerge from this structure. As Rovelli argued in 2011: 'A theory from which we cannot compute is not a good theory'.[9] He went on to explain that the difficulty of computing anything at all in quantum gravity is 'nearly mythical'.

Exploring how the theorists sought to recover physics from LQG requires us to dig a little deeper into the quantum field theory formalism developed by Feynman and to become a little more familiar with some of its terminology. In *QED: The Strange Theory of Light and Matter*, first published in 1985, Feynman explained that to understand pretty much all the phenomena associated with light (photons) and electrons, we need to understand only three basic actions:[10]

1. A photon goes from place to place.
2. An electron goes from place to place.
3. An electron emits or absorbs a photon.

That's it. We can represent these three actions in a simple Feynman diagram. If you think this all seems deceptively simple, and therefore too good to be true, you'd be right. Behind this simple diagram sits a bunch of mathematics which collectively form the *equations of motion* for the photon and electron. From these equations it is possible to calculate the *propagators* for each action. These govern the probability (given as the square of the probability *amplitude*) for the electron to move from here to there (represented as a straight line) and for the photon to move from someplace else to there (wavy line). There's also mathematics for calculating the strength of the interaction (or *coupling*) at the vertex where the electron 'meets' the photon and absorbs or emits it. There are some very specific rules—called Feynman rules—for setting up the equations and performing the calculations.

Let's make this a little more specific. We'll simplify things by assuming that the electron and photon are constrained by the equations to move only in one dimension, which we'll assume is the x-direction. So, the electron moves from here (position x_1 at a time t_1) to there (position x_2 at a time t_2). The propagator then contains all the information that we need to tell us how the electron moves from $x_1 t_1$ to $x_2 t_2$.

You can now begin to sense the nature of the problem. In LQG there is no 'here' and 'there'. There is no 'place to place', no ruler against which distances along the x-direction can be measured (and, indeed, no 'x-direction'), and no clock to tell us how long

this might take. The axes of Feynman diagrams are very deliberately labelled 'space' and 'time' because of course the approach
assumes a background spacetime.

But Feynman's approach is second nature not only to particle
theorists but also experimentalists and, as such, it cannot be
ignored. Somehow, a theoretical framework whose foundations
are built on the assumption of a background spacetime had to be
connected with a theory in which spacetime itself is emergent. Like
circus acrobats performing without a safety net, the LQG theorists
had to learn how to do particle physics without spacetime.

In a book like this it's probably okay for me to wave my hands in
the air and talk in a very general way about 'space' and 'time', provided you get the gist. But practical calculations of the type derived
from Feynman diagrams need to be a little less hand-waving and a
lot more specific. Most importantly, we need to acknowledge that
the physics we're describing takes place within a finite volume of
space and in a finite time. This means drawing a *boundary* around
the region of spacetime in which all the interesting stuff is going
on and forgetting about anything that's going on outside this.

When we do this in quantum field theory, we find that it
becomes necessary to add a 'boundary term' to the expression
for the action in order to get the correct equations of motion.

Before we go on to see how this helps, there are a couple of
seemingly trivial observations I'd like to make with the aid of
some high school geometry. Think of a circle. We know from
Euclidean geometry that the circle is a 2-dimensional object with
an area A given by πr^2, where r is the radius. The property of
'2-dimensionality' is manifested by the area, which is measured,
for example, in square centimetres (a circle with a radius of
5 centimetres has an area of about 78.5 square centimetres, or cm^2).
But the boundary is 1-dimensional—the circumference of the
circle is a line whose length is given by $2\pi r$, or 31.4 centimetres.

Likewise, a cube is a 3-dimensional object, reflected in the volume which is given by the cube of the length, l. A cube with $l = 5$ centimetres then has a volume V of 125 cubic centimetres, or cm^3. But the boundary of the cube is 2-dimensional. It is formed by six sides of equal area, giving a total area of $6l^2 = 150$ cm^2. Generally speaking (and rather obviously), the boundary of any region has one less spatial dimension than the region it encloses.

The second thing to note is that the information contained in the expression for the boundary gives us everything we need to calculate whatever lies within it. In the case of the circle, if we know the circumference we can calculate the radius by dividing this by 2π, and we can then use this to calculate the area. Alternatively, we can say that the circumference is $\sqrt{4\pi A}$—the area is, in a sense, *encoded* in this expression for the boundary circumference. Likewise, the area of the cube is $6\sqrt[3]{V^2}$: the volume of the cube is encoded in this expression for its boundary area.

These are obviously extremely simple examples from elementary geometry, and the problem becomes a lot trickier as the shape of the enclosed region becomes more complicated. But I think you get the point.

Now in quantum field theory the boundary term that must be added to the equation for the action has one less dimension than the region it encloses. It also *encodes* information about the physics going on inside. If we can then specify the initial conditions of the system, we have all the information we need to calculate the *classical* solutions.

Smolin had introduced a finite boundary in LQG in his 1995 'linking' paper, but he had not developed this into a formulation that could be used to perform calculations. It was therefore Rovelli's colleague Robert Oeckl, at the Centre for Theoretical Physics in Marseille, who went on to figure out the principles of a 'general boundary' formulation of quantum field theory in 2003.

This is a much more elaborate 'encoding' than the simple geometric examples I gave earlier. Oeckl likened the approach to holography, writing: 'In this sense the formulation is "holographic", i.e. the information about the interior of a region is encoded through states on a boundary'.[11] This is a good analogy. In a hologram, all the information needed to reconstruct a three-dimensional image is encoded in a two-dimensional interference pattern recorded on a piece of photoresist or photopolymer.

Oeckl and Rovelli had many discussions about how to compute things in LQG. Rovelli vividly recalls one such conversation which prompted the kind of flash of inspiration that is sometimes referred to as a 'light bulb moment'.

The information encoded on the boundary represents the sum over all the different histories of the quantum physics going on inside. For example, if the action we're interested in involves an electron moving from place to place, then the information encoded on the boundary around the spacetime region in which this occurs takes account of *all* the different ways this can happen, including histories in which virtual particles are created and destroyed. If we choose a large enough region, and specify the initial conditions, the physics on the boundary tends towards a *classical* trajectory, the result of constructive interference between all the different quantum pathways, which are confined to the interior.

Now the above discussion is focused on the idea of a boundary around a region of spacetime, but the mathematical structure does not depend on the presumption of a spacetime background. *It applies equally to a boundary around a region of spinfoam.*

Rovelli became very excited. Over the course of the next couple of years, he and a number of colleagues from various institutions including the Universities of Rome, Torino, and Pisa, Penn State University, and the Perimeter Institute worked to apply these techniques to a spinfoam description.

Instead of histories involving the creation and destruction of particles, the histories in the spinfoam formalism relate to the creation and destruction of quanta of space. By choosing a suitably large-scale region (enclosing many quanta of space and many transitions between spin networks), the 'holographic physics' encoded on the boundary determines the geometry of the boundary and tends towards a limit we identify with physics occurring in a continuous spacetime.

The boundary approach now gave the theorists the ability to *calculate* the actions of particles without the need to assume a background spacetime. The result—a particle moving from place to place *in* spacetime—appears in the equations for the boundary itself. 'It was amazing for me that this could be done', Rovelli said.[12] Oeckl had helped to remove a stumbling block that had long been in the way.

It was a very neat trick, but now the theorists somehow needed to discover whether it could be used to yield a correct description of the physics. They had not yet introduced matter into this picture—this was still a theory of space and time only—but as we've seen this doesn't preclude the possibility of massless particles. The physics they needed to describe was that of the graviton.

Now, it's all very well to deduce an expression for the graviton propagator—describing the action of a graviton moving from place to place—but, after all, the graviton is a hypothetical particle which, as we saw earlier, is very unlikely ever to be detected in our universe. So, how were the theorists supposed to know whether they had the 'correct' description?

We know that the covariant approach to quantum gravity presents great problems of renormalization, but in a low-energy approximation the divergent terms which cause the equations to blow up can be quietly ignored. This allows us to deduce an approximate, low-energy expression for the graviton propagator based on the 'traditional' approach favoured by particle theorists. Rovelli and his colleagues were able to show that their boundary approach applied to the spinfoam formalism enabled the calculation of a graviton propagator that is entirely consistent with that deduced from the low-energy limit of covariant quantum gravity.

What this says is that two different theoretical approaches—which, of course, may or may not provide correct descriptions of reality—produce results that are at least consistent with each other. But there was a little more. In the large distance limit, the graviton propagator calculated from LQG was found to depend on the inverse square of the distance between 'here' and 'there'. It also depends on physical constants, such as Newton's gravitational constant G, in a way that is entirely consistent with the classical limit.

Although a logical connection had been established, this was hardly a straight line that begins with quanta of area and volume and ends with our classical intuition for the reason why things fall down. The issue concerns the nature of the comparison. In quantum mechanics, it is technically possible to extrapolate quantum behaviour to a limit, typically involving large quantum numbers, and so recover classical physics. This is called the *correspondence principle*. Unfortunately, establishing a correspondence between LQG and general relativity and, ultimately, Newton's law of universal gravitation is not simply a matter of extrapolating to large numbers of the quanta of space. The mathematical derivations of the graviton propagator are complex, requiring many assumptions, and there remain many open questions.

Nevertheless, Rovelli was sufficiently emboldened to remark, at least informally: 'We have calculated Newton's law starting from a world with no space and no time'.[13]

Feynman had been famously critical of the string theory community, claiming that he didn't like the fact that the theorists weren't calculating anything.[14] There could be no doubting that the real achievement of the LQG theorists was their use of the boundary approach to perform some physically meaningful calculations:[15]

> In our opinion, the interest of the calculation ... [lies] in the fact that it shows how some low-energy quantities with a transparent physical meaning can be computed, starting from the abstract context of a background independent formalism.

It didn't matter that the graviton might never be observed. The theorists had shown how it was possible in principle to do particle physics without the safety net of a background spacetime.

They had also learned a few new things along the way. The original spinfoam model developed by Barrett and Crane was found to be in need of modification, leading to the introduction of a new parameter. This is called the Immirzi or Barbero–Immirzi parameter, usually given the symbol γ and named for Italian theorist Giorgio Immirzi and Spanish theorist Fernando Barbero. We will meet this parameter again in our discussion of black hole entropy in Chapter 15, so I won't say too much about it here. The parameter modifies the expression for the spectrum of area, and so changes the size of the quantum of area somewhat.[16] Suffice to say, its appearance is a bit of a puzzle and suggests that there is some important aspect of the theory that is not yet completely understood.

Nevertheless, the theorists had demonstrated the possibility of doing particle physics in a theory without spacetime. Now it was time to get serious about *matter*.

12

FERMIONS, EMERGENT PARTICLES, AND THE NATURE OF STUFF

Loop quantum gravity was never really intended as a candidate theory of everything. If the theorists could show that it provides a viable quantum theory of gravity, then it could be argued that it should take its place *alongside* the existing quantum field theories used so successfully to describe subatomic and subnuclear physics.

That's quite easy to say, but it's not at all clear what this would mean, or how it might work. LQG is simply not like the quantum field theories of the standard model. The latter describe the dynamics of matter and force particles against an assumed background spacetime. The former provides a description of an emergent spacetime, so there is a sense in which LQG eventually *has* to at least touch the standard model theories. Like LQG itself, these should arguably be reformulated without the spacetime safety net—even though doing this might not necessarily serve to unify them. But hey, who knows?

One potential benefit is that by formulating them on spinfoam rather than in spacetime, the quantum field theories themselves might become more 'orderly'. It has long been assumed that successfully quantizing spacetime would help to 'regulate' theories such as QED, thus completely avoiding the need for renormalization.

To get some sense for why this might be it will help to recall a few things that we learned in Chapter 4. Some terms in the equations of QED mushroom to infinity because the electron interacts with its own, self-generated electromagnetic field. Renormalization is then simply a mathematical fix which gets all the infinite terms to cancel out, leaving only physically sensible predictions. But it would obviously be much better if the infinities didn't arise in the first place. After all, infinity is a purely mathematical concept—*infinity does not exist in the real world*. When infinities start to jump out of the equations they're telling us that, in some essential respects, our mathematical description is not properly representing reality.

This kind of problem has a long history. The ancient Greek philosopher Zeno of Elea devised a famous paradox in which Achilles runs a race against a tortoise. It's clear that Achilles and the tortoise are unevenly matched, but our hero has a strong sense of honour and fair play, and he agrees to give the tortoise a head start. So, Achilles waits until the tortoise has reached a certain position—halfway to the finish line—before he sets off. He seems certain to win, but by the time he has reached halfway, logic suggests that the tortoise will inevitably have moved on a certain additional distance. By the time Achilles has covered this additional distance, the tortoise will have moved on a little further still. We can go on like this for ever, it seems. Each time Achilles reaches the point where the tortoise was, the tortoise has moved a little further ahead. It seems that Achilles will never overtake the tortoise.

At the heart of Zeno's paradox lies the seemingly innocent observation that a continuous line can be divided into an infinite number of points. But if there's an infinity of points between Achilles and the tortoise, then no matter how fast Achilles moves, there's clearly no way he can catch up in a *finite* time.

It's not too difficult to find a resolution. Whilst it is correct to suggest that a continuous line can be *mathematically* divided into an infinity of points, this does not mean that a distance or an area or a volume in the real world can be *physically* so divided. There is no such thing as infinity in the real world.*

This leads us directly to the notion that the material world is not continuous and endlessly divisible, but is instead composed of discrete atoms. The principal lesson of physics from the very beginning of the twentieth century is that every property we had previously judged to be continuous—matter, radiation and energy—turned out instead to be discrete. The stuff of the universe comes in 'lumps' and 'bits'.

If we push this to its logical end point, we conclude that space (or spacetime) is itself likewise not continuous or endlessly divisible. Perhaps it is only when we *assume* that a background spacetime is continuous that infinities start jumping out of the resulting quantum field equations, demanding elaborate renormalization techniques to get rid of them (or failing to do so, as the case may be).

And thus we're led to the *real* difference between the premises of string theory and LQG. String theory is essentially a *particle*

---

* The paradox can also be 'resolved' by noting that the incremental distances that Achilles must make up get successively smaller and smaller, and we know that an infinite series of such increments will sum to a finite result, implying that he *can* cover the distance in a finite time. But this is a purely mathematical argument—we compensate the 'unnatural' mathematical division into an infinity of points by introducing a similarly 'unnatural' infinite sum over these points.

theory—it seeks to eliminate all the troublesome infinities by assuming that particles are *not* infinitesimally small point-masses but rather extended two-dimensional or multidimensional objects. The result is a theory of strings *in* a background of continuous spacetime which requires extra hidden dimensions and, to make them less unruly, the additional assumption of supersymmetry. This eliminates the point-mass problem, but not the problem posed by the assumption of a continuous spacetime. In the string universe, Achilles will still struggle to catch up with the tortoise.

In contrast, LQG is a theory of spacetime *geometry*—it seeks to eliminate the infinities by combining general relativity and quantum mechanics, showing that spacetime is discrete, rather than continuous.

This is okay as far as it goes, but there's obviously much more to nature than spacetime geometry (and massless gravitons). Reality tends to have a lot more going on in it. The challenge then is to find a way to incorporate particles with mass into this picture. And, as all matter is composed of fermions (quarks, leptons), this means finding a way to couple spinfoam LQG with *fermion fields*.

Lee Smolin, Carlo Rovelli, and their colleagues had already developed ideas for how to incorporate fermions in LQG back in the 1980s. But the success of the boundary approach in the calculation of the graviton propagator led the theorists to search more intently for methods to link the quantum fields of fermions to a spinfoam description. Attempts had been made to do this before, but the field was now sufficiently mature to suggest that the time was right to try again.

The theorists were under no illusions. Their principal concern was to get a sense that the coupling of quantum gravity to fermions could, in fact, be done. It was not yet clear that the large-scale limit of a theory of spinfoam-coupled fermions would be equivalent to the corresponding standard model quantum field theory. In other words, the theorists couldn't be sure that a theory of spin-foam-coupled electrons, for example, would yield QED in the theory's large-scale limit. 'Are we capable of doing so entirely at present?' asks Rovelli, rhetorically. 'No, not entirely. We have numerous indications that the limit should come out right. Still, it is a cumbersome and intricate situation'.[1]

And even if the limit does eventually come out right, perhaps there is little to be gained by seeking to use spinfoam LQG to reproduce already well-established results, such as the QED prediction of the g-factor for the electron. As Rovelli explains, 'nobody would claim that it would make sense to test general relativity by recalculating the stability of the Brooklyn Bridge, because the result would be the same as with Newtonian mechanics'.[2]

Whilst this is certainly true, it's important to note that general relativity can be demonstrated to reduce to Newton's gravity in flat spacetimes at low speeds—this is Einstein's principle of consistency. A similar principle of consistency for spinfoam LQG and QED is conspicuous by its absence. I guess there is always the *hope* that pushing a theory whose domain of applicability lies at the Planck scale into the subatomic and subnuclear scales of quantum field theory might yield some new *explanatory* insights—some new ways of thinking about the behaviour of quarks and electrons in a discrete, emergent spacetime, even though the results might be the same. But, in truth, the seventeen-orders-of-magnitude difference in these scales suggests that this hope, if it really exists at all, is indeed very faint.

Rovelli, working with Eugenio Bianchi, Muxin Han, and Wolfgang Wieland at the Centre for Theoretical Physics in Marseille and Elena Magliaro and Claudio Perini at Penn State, figured out how to attach fermion fields to spinfoam in a paper published in 2013. It was 'surprisingly simple'.[3] To a certain extent, the theorists simply copied the way fermions are coupled in lattice QCD, taking advantage of the fact that, at the boundary, the spin-foam tends towards a dynamic lattice of spacetime points. The challenge was to find the best way to write the fermion action—a fermion going from place to place—and they succeeded in developing a technical trick for doing this.

The result is a quantum theory of gravity *interacting with fermions*.

The properties of fermions are governed by something called CPT symmetry. Here, 'C' stands for charge conjugation. If the physics of a particle is invariant under C-symmetry transformations this means that replacing its electrical charge with its opposite (or, alternatively, replacing the particle with its antiparticle) makes no difference. 'T' stands for time reversal—replacing t in the equations describing the dynamics of a particle with $-t$. Although it is our experience that time progresses only ever in one direction, the equations of physics are generally indifferent—they apply equally to events unfolding forwards or backwards in time (we say that the physics is invariant under T-symmetry transformations).

'P' stands for parity transformation or parity inversion, and requires a little more explanation, for which we need to switch from a particle to a wave description. A sine wave moves up and down as it oscillates between peak and trough. Parts of the wavefunction have positive amplitude (as it rises to its peak and falls back) and parts have negative amplitude (as it dips below the axis heading for the trough and comes back up again). The parity of the wavefunction describes its behaviour as we change the signs of the spatial coordinates in which the wave propagates. We can think of this as changing left for right or up for down or front for back. Changing the signs of all three coordinates simultaneously is then a bit like reflecting the wavefunction in a special kind of mirror which also inverts the image and its perspective. The image is inverted left to right and up to down, and front goes to the back as the back is brought forward to the front.

If reflecting the wavefunction in such a 'parity mirror' doesn't change the sign of the amplitude, then the wavefunction is said to possess even parity. If the amplitude does change sign (from positive to negative or negative to positive), then the wavefunction is said to have odd parity.

Like spin, parity is a property without many analogies in classical physics that are not thoroughly misleading. It is closely connected with and governs angular momentum in elementary particle interactions. As far as the physicists of the late 1950s could tell, in all types of interactions parity is something that is conserved, like angular momentum itself. In other words, if we start with particles which when taken together have overall even parity, then we would expect that the particles that result from some physical interaction would also possess overall even parity. Likewise for particles with overall odd parity.

This seemed consistent with the physicists' instincts. How could it be possible for the immutable laws of nature to favour such seemingly human conventions of left versus right, up versus down, front versus back? Recall that, as he developed his spin network formulation, Roger Penrose had asked himself how an electron is supposed to know which way is 'up'. Surely, no natural force could be expected to display preferences based on notions of right- or left-handedness?

As reasonable as this seems, in fact this is not consistent with what we observe. Parity is conserved in all electromagnetic, gravitational, and strong colour force interactions. But experiments performed in the mid–late 1950s demonstrated that nature does indeed exhibit a peculiar 'handedness' in interactions

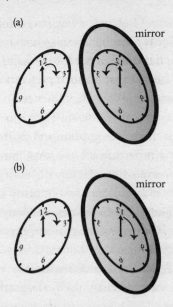

Figure 22. Reflecting an object in a 'parity' mirror may produce one of two results. In (a), the clock on the left is mirrored exactly by the reflection on the right, and parity is said to be preserved. In (b), the reflection does not mirror the clock and parity is said to be violated.

involving the weak force (see Figure 22). In fact, by convention, only 'left-handed' particles and 'right-handed' antiparticles actually undergo weak force interactions.*

These results came as a shock to many of the older generation of theorists, such as Pauli. Just days before receiving news of the experiments he had written: 'I do *not* believe that the Lord is a weak left-hander, and I am ready to bet a very large sum that the experiments will give symmetric results'.[4] It was fortunate that

---

* In beta-radioactive decay, a down-quark transforms into an up-quark (turning a neutron into a proton), with the emission of a W⁻ particle. This further decays into a 'left-handed' electron and 'right-handed' electron antineutrino.

he did not make a bet because, as he later admitted: 'It would have resulted in a heavy loss of money (which I cannot afford)'.[5]

It wasn't so much the preference for left-handedness that baffled Pauli; it was the fact that no such preference is apparent in strong force interactions. Why the difference?

Such 'handedness' is called *chirality*. For a time it was believed that although both charge conjugation and parity symmetry can be violated in certain particle interactions, joint CP symmetry would still be conserved. But experiments performed in 1964 at Brookhaven National Laboratory by Princeton physicists James Cronin and Val Fitch showed that in certain decay processes of neutral kaons (mesons formed from mixtures of down/antistrange and strange/antidown quarks) CP-symmetry is violated, too. Only when we add time-reversal symmetry does nature once more start to behave itself. The combination CPT is an exact symmetry.

Nobody really understands why this happens.

Han and Rovelli were able to show that when coupled to a spinfoam, the fermion propagator is, as required, invariant under CPT symmetry. They found that the boundary behaviour could be represented in simple Feynman diagrams embedded in the spinfoam amplitudes. They had no reason to doubt that the physics of interacting fermion fields would be represented in much the same way. It seemed that fermions, hitherto considered only as quantum fields existing in spacetime, were quite at home on a sea of spinfoam.

But there was still a problem. Sen and Ashtekar had chosen the original spin variables to reflect the property of chirality, and so there is no issue in principle in introducing chiral fermions into spinfoam LQG. But as soon as fermion fields are fastened to a lattice, they become rather badly behaved, exhibiting an unnatural *fermion-doubling*. Spurious mirror-image states appear, with every original left-handed fermion breeding a right-handed counter-part, along every spatial dimension.

This is a problem also in lattice QCD, governed by a powerful 'no go' theorem, named for Danish theorist Holger Nielsen and Japanese theorist Masao Ninomiya. This states that free fermions modelled on a discrete lattice can only yield consistent results when coupled equally to both left-handed and right-handed variants. But the weak interaction explicitly favours left-handed fermions, so the coupling cannot be equal. In lattice QCD various 'work-arounds' have been introduced to get rid of the extra fermions—one of which, for example, artificially breaks the chiral symmetry, pushing the doubled particles into a high mass range so that they no longer interact and get in the way.

Han and Rovelli considered some of these work-arounds and rejected others. Despite the problems they remained optimistic. 'It is not hopeless in our opinion for the following reasons: the summing over all the geometries in the spinfoam model make it hopeful that the fermion doublers are cancelled in a similar way to those on a random lattice'.[6] Smolin doesn't agree, arguing that fermion doubling also occurs on a random lattice.

Working with Jacob Barnett at the Perimeter Institute, in 2015 Smolin explored fermion doubling in some detail, demonstrating that the original canonical LQG suffers unavoidably from this problem and so cannot be used to describe chiral fermions.[7] He believes this is a major obstacle to the viability of LQG theories as descriptions of nature. 'We have some ideas as to how this problem might be overcome in spinfoam models,' he explained, 'but haven't worked it through yet'.[8] It remains an open question.

These kinds of problems pull the theorists up short, forcing them to stop and think about the mathematical apparatus they're using. But they also create niggling doubts that we really don't yet understand the nature of particles at all. Are all these difficulties bred from mathematical technicalities or are they really trying to tell us something rather deeper and more profound about the nature of material substance?

Remember that LQG is constructed from interconnecting loops of force which form spin networks, with the nodes representing the points of intersection where quanta of volume reside and the links representing the surfaces between adjacent nodes which can be characterized as quanta of area. If we return to the idea of a space formed from closed loops, then fermions are simply *open loops*, characterized by a spin quantum number of ½. As Smolin explains, 'some of the more exciting moments in the early days were discussions about how fermions would emerge from free ends or punctures of spin networks'.[9]

If the lines of force do not form closed loops, then they must start and finish at an electrical charge (positive or negative), or a colour charge (red, green, or blue). Now this may sound somewhat familiar: in early versions of string theory, matter particles are represented as open strings. And, indeed, insofar as string theory and LQG are built up from models based on lines of force, then they clearly start from broadly the same point. It's where they go next and what they assume that makes them so very different.

The open loops appear as an extra spin-½ 'face' on a spin network, tying matter particles very firmly onto the fabric of an emergent spacetime. This is all very reminiscent of some speculative theorizing by Wheeler and Misner, who in 1957 suggested that charged particles might be no more than lines of electrical force threaded through Planck-scale wormholes, hypothetical 'tunnels' connecting different points in spacetime.* They imagined

* Wormholes are hypothetical because they have never been seen (or, more correctly, there are no observed physical effects that can be unambiguously ascribed to wormholes). But they nevertheless derive from valid solutions of Einstein's field equations of general relativity.

that the lines of force 'disappear' into one wormhole and 'reappear' from another. The force lines form closed loops, but because they disappear 'here' and reappear 'there' they look to all the world like open loops. In other words, through this assumption we've conjured matter particles from gravity alone.

Wheeler called it an 'already unified theory'.[10] Rather than unifying gravity and electromagnetism—as Einstein had spent many years in futile attempt—why not model the electron as a fundamental property of spacetime itself? Now surely that would be very neat.

Smolin showed in 1994 that, broadly speaking, a theory which couples general relativity to fermions is entirely equivalent to general relativity with no matter but in which certain kinds of 'minimalist' wormhole are permitted. As far as the physics is concerned, electrons and wormholes are one and the same. This conclusion was apparent in the later work on spinfoam fermions. In 2013, Rovelli and his colleagues noted that, at fixed times, the fermion loop is non-local—it disappears outside spacetime to reappear at some distant point.

This may sound bizarre, but the relationship between fermions and gravitational objects runs even deeper. I mentioned in Chapter 2 that German mathematician Karl Schwarzschild worked out a set of solutions to Einstein's gravitational field equations for the case of a large, electrically uncharged, non-rotating spherical body. These solutions admit the possibility of black holes, called Schwarzschild black holes. There are now known to be four types of black hole solution. One of these applies to the case of spinning, electrically charged bodies and is named for New Zealand mathematician Roy Kerr and American physicist Ted Newman.

Here's the thing. A Kerr–Newman black hole has a g-factor exactly equal to 2, equivalent to that of the 'Dirac electron' before radiative corrections are applied in QED. Now we tend to think of black holes as resulting from the gravitational collapse of large astronomical objects such as stars, but *microscopic* black holes are theoretically possible.* There are all sorts of reasons why electrons-as-Kerr–Newman black holes might not be very feasible, but this is exactly the kind of coincidence that has the theorists reaching for their thinking caps. Some have argued that the coincidence reflects 'a deep common root of quantum theory and general relativity'.[11]

However it's done, conjuring fermions from lines of gravitational force is, of course, a long way from explaining the three generations of matter particles that, together with the force particles and the Higgs boson, collectively form the standard model of particle physics. But when Smolin chanced on a preprint by a young Australian theorist based at the University of Adelaide called Sundance Bilson-Thompson, he sensed that it might indeed be possible to construct a theory in which the particles of the standard model *emerge* from the twisting and braiding of spin networks.

Now, Smolin continues to be fully committed to LQG and the exploration of its potential to make testable predictions, but his restless energy and insatiable appetite for new ideas will often lead him into more speculative territory. He is, of course, interested in the detailed nuts and bolts required to turn LQG into a predictive theory, and his commitment to the fundamentally important roles of theoretical prediction and experimental testing that lie at the core of the scientific method is unquestioned.

---

* Fears were raised before the Large Hadron Collider was first switched on in 2009 that high-energy proton–proton collisions would generate microscopic black holes, heralding the end of the world. As we've seen, the LHC has operated for long periods over the past seven years or so and, as far as I'm aware, the world has not ended.

The near-simultaneous 2006 publication of his book *The Trouble with Physics* and Columbia University mathematical physicist Peter Woit's *Not Even Wrong*—both highly critical of the superstring theory programme and the culture in which it is embedded—triggered what was to become known as the 'String Wars'. Smolin is naturally averse to conflict and confrontation, but he was drawn inexorably into a series of debates about the nature of science and the future of theoretical physics.[12]

But at the same time Smolin is also rather more concerned to explore what a theory describing space in terms of elementary quanta might have to say about the nature of physical reality at its deepest level. He is not averse to metaphysics per se and he's more than willing to jump in where others might be more hesitant. The accepted theories that frame what we might think of as the 'authorized' description of physical reality—quantum theory and general relativity—arguably began as metaphysical speculations. But they were speculations which implied empirically observable consequences, and this is what makes them scientific.

As the evidence accumulated for the existence of several generations of quarks, it was perhaps inevitable that a few theorists would want to explore the possibility that these might be, in turn, composites of something even more elementary. There was also an incentive to find ways in which the very distinct families of quarks and leptons might be reconciled, to be seen as composites of the same set of material building blocks. Such hypothetical entities became known generally as 'pre-quarks', or 'preons'.

In 1979, Haim Harari at the Stanford Linear Accelerator Center in California and, independently, Michael Shupe at the University of Illinois at Urbana-Champaign published details of a preon model consisting of just two particles and two antiparticles. These possess spin ½ and electrical charges of 0 and ⅓, which are then used to build the first generation of quarks (up, down) and leptons (electron, electron neutrino) and their antiparticles. The second and third generations of quarks and leptons were simply interpreted as higher-energy (and so higher-mass) states of the first generation.

Harari labelled the charged preon T and the neutral preon V (drawing on the Hebrew version of Genesis, in which the world emerged from 'Tohu va-Vohu'—unformed and void).[13] The three different colours of the up quark, with net charge $+\frac{2}{3}$, then result from the combinations $TT\overline{V}$, $T\overline{V}T$, and $\overline{V}TT$, and the electron from the combination $\overline{TTT}$, where $\overline{T}$ is the antiparticle of T. As the leptons are formed from only one kind of preon (either TTT or VVV or the antiparticle combinations) they do not exhibit colour.

It is possible in this scheme to account for force particles, too. Gluons change the order of the preons in the quarks, for example turning TTV into TVT, so changing the quark colour (from red to green, say). The W and Z bosons are assumed to be composites of the electron and neutrino and their antiparticles. For example, if the W⁺ is taken to be the combination $TTT\overline{VVV}$ (a positron and neutrino), then its action on a down quark such as $\overline{VVT}$ is to turn it into an up quark, TTV.*

This might seem all rather contrived, and indeed it is. But arguably it's no more contrived that the original quark model, with all its flavour, colour, and astonishing fractional electric charges.

Bilson-Thompson translated the Harari-Shupe scheme, which was based on the idea of point-particles, into a topographical 'toy model' based instead on a series of graphs in which three 'ribbons' are braided in different ways. A ribbon can be charged (corresponding to Harari's T) or it can be neutral (V). Force-carrying bosons are unbraided. Second and third generation

---

* Remember, particles and their antiparticles mutually annihilate.

particles simply have more complex braids than the first generation. The result is simpler than the original Harari–Shupe model and has the additional benefit of providing a basis for chirality, through braids that twist to the left or to the right, forming mirror images.

Bilson-Thompson offered no suggestions regarding the nature or origin of the ribbons themselves, though he 'toyed with the idea of them being micro-wormholes, which wrapped around each other'.[14]

'As soon as I read this paper,' Smolin later acknowledged, 'I knew this was the missing idea, because the braids Bilson-Thompson studied could all occur in loop quantum gravity'.[15] Recall from Chapter 8 that, one sunny day in Santa Barbara in February 1986, Smolin and Jacobson had found the first few exact solutions to a spin-connection reformulation of the field equations of general relativity. Smolin knew that Bilson-Thompson's braided ribbons could all be admitted as exact solutions to these equations, suggesting an astonishing hypothesis. *All the particles of the standard model represent different ways of twisting and braiding gravitational lines of force.*

'We knew about braiding in 1987,' Smolin admitted, 'but we didn't know it corresponded to anything physical'.[16] Some years later, following his work on the 'linking' paper and collaborations with Markopoulou, Smolin had a much better idea how to get from loops to ribbons which could then braid and twist.

Conjuring particles from spacetime itself would be the ultimate realization of both Wheeler's 'already unified theory' and Kelvin's nineteenth-century vision of atoms as 'knots in the ether'. But could it be done? Smolin and Markopoulou invited Bilson-Thompson to collaborate, and they developed an elaboration of the model in which the three 'ribbons' became three links in a spin network connected to a common set of nodes.[17] The electrical charges that Bilson-Thompson had introduced in his model could be replaced by twisting the links one way (⅓ positive charge) or the other way (⅓ negative charge).

Twisting two of the three ribbons forming an up quark then gives a total ⅔ positive charge. As there are three ways this can be done (we can twist the first and second ribbons, the first and third ribbons, or the second and third ribbons), there are three different kinds of up quark corresponding to the three quark colours. A similar logic applies to the down quark. Because all three ribbons are twisted in an electron (to give a total charge of −1), different colours are not possible. Uncharged particles, such as the electron neutrino and antineutrino, have no twists.

The ribbons can also be braided. Braids which thread first under then over from left to right represent left-handed particles. Threading under then over from right to left represents right-handed particles. These are not mirror images, and Smolin was hopeful that with this construction the theorists could avoid the fermion-doubling problem.

A series of 'braid states' with a minimum of two 'crossings' and two different types of twist is sufficient to account for the particles that make up the first generation of standard model particles. The patterns for the up quark, down quark, and electron are illustrated in Figure 23a, with Bilson-Thompson's original graphs on the left and the quantum gravity-derived graphs on the right. Only quarks of a single colour are shown, and all particles are left-handed.

Because of the way the braid states are constructed, only a left-handed electron neutrino and right-handed electron antineutrino are possible, consistent with observed particle interactions—see Figure 23b. The theorists speculated that

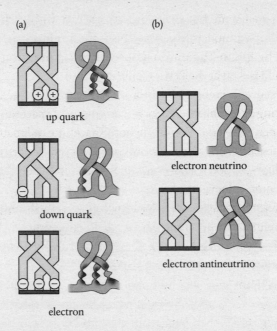

(a)   (b)

up quark

down quark

electron

electron neutrino

electron antineutrino

Figure 23. Smolin and Markopoulou helped to turn Bilson-Thompson's topographical toy model, shown on the left in both (a) and (b), into a system involving the braiding and twisting of gravitational lines of force in LQG, shown on the right.

second- and third-generation particles would then involve similar braid structures with more crossings.

Of course, there were still lots of things that this kind of scheme could not explain, such as the masses of the particles, and initial excitement quickly gave way to a more sober assessment. Speculations about the second and third generations of standard model particles proved to be unsubstantiated—no simple correspondence with braid structures could be established. There is also a sense in which the first-generation braid structures are too stable—the emergent particles may propagate through spacetime but they couldn't interact with each other. More complex models have since been studied in an attempt to fix this, by Bilson-Thompson, Louis Kauffman in Chicago, and Jonathan Hackett at the Perimeter Institute.

It was always going to be a long shot. After all, LQG was far from being an established theory of an emergent spacetime, with many open problems. Conjuring both spacetime *and* matter and force particles from spin networks remains a rather fascinating possibility—a good trick if you can pull it off in a way that offers some genuinely new insights.

But, on present evidence at least, it requires rather more magic than can normally be admitted in a scientific theory.

13

RELATIONAL QUANTUM MECHANICS AND WHY 'HERE' MIGHT ACTUALLY BE 'OVER THERE'

The scientific collaboration between Carlo Rovelli and Lee Smolin now spans thirty years. They are also close friends. But there are two subjects on which they profoundly disagree. The second is the nature of time, to which we'll return in Chapter 16. The first is the interpretation of quantum mechanics.

Now, disagreement is not necessarily a bad thing, as Rovelli explains: 'Lee and I have constantly exchanged ideas and have always had a curious scientific relationship where we have found ourselves often in disagreement about issues, and in spite of that, or maybe precisely because of that, we have kept learning from one another and influencing one another I believe far more than we ourselves realized'.[1] Smolin concurs: 'With all our disagreements I have never felt a strain on our friendship. Quite the opposite: if we agree about everything then one of us is redundant. I think we have always listened to each other'.[2]

As we've seen, Einstein was rather uneasy about what quantum theory seemed to be suggesting about the nature of physical reality. He argued that apparent contradictions between a microscopic world of quantum probability and uncertainty and the more common-sense classical world meant that quantum theory is in some fundamental sense incomplete.* All the different possible extensions and interpretations of quantum theory that have been devised since Einstein's famous debate with Bohr are essentially attempts to answer the 'completeness question'.

Ever since he read the original EPR paper as a first-year college student, Smolin has felt uneasy about the interpretation of quantum theory. Like Einstein, Schrödinger, Bell, Leggett, and—as we will soon see—Penrose, he is inclined to believe that the theory must be incomplete in some essential way. Something is missing.

Rovelli had always found quantum theory to be profoundly revolutionary, but not inconsistent or incomplete in any sense. A brief period of experimentation with LSD as a student had helped to convince him that reality is very different from perception: 'Psychedelic drugs, I think, have an impact on you because it's like a version of the meditation experience. You see the world in a totally different way, you realise your perception is limited'.[3] He did not consider quantum mechanics to be 'a problem'.[4]

---

* I'm not aware of any universally agreed definition of what constitutes 'common sense'. Remember that the general theory of relativity is a theory of classical (non-quantum) physics, and for most people general relativity is hardly common sense. In this context, by common sense I mean the logical structure of classical physics which is familiar from everyday experience. This follows that, and when one happens, we expect the other, with unfailing predictability and certainty. Classical objects are 'local', and don't form weird superpositions. They continue to possess the properties we ascribe to them even when we're not looking.

Smolin has a vivid memory of a three-way brainstorming session with Rovelli and Louis Crane, when they were all together at Yale. They tossed ideas around about the relation between elements of reality and the observer, the philosophy of Gottfried Leibniz, and quantum mechanics. 'The way I remember it,' Smolin explained, 'We each took the basic ideas and constructed theories which expressed the idea that quantum mechanics was relational'.[5]

Rovelli built on these discussions and used them to sharpen his own distinct interpretation. Like the physicists of the 'Copenhagen school'—Bohr, Heisenberg, and Pauli—and philosophers such as Bas van Fraassen and Michel Bitbol, he assembled a logical structure that he believes clearly demonstrates that quantum theory is quite complete. There is nothing more to add.

What needs to change is the way we think about what it's telling us.

Rovelli's interpretation was first presented in a couple of papers published in the late 1990s. He subsequently refined and further sharpened his views—collectively referred to as *relational quantum mechanics*—in collaboration with Federico Laudisa and Matteo Smerlak in 2005–7. The basis for this interpretation can be found in the role of the 'observer' in both relativity and quantum theory.

Remember that in his theories of relativity Einstein was at pains to put the observer firmly back into the picture, making measurements with rulers and clocks and performing observations *inside* the physical reality that is being examined. As we've seen, the result was a dramatic change in the status of space and time, which he dislodged from their grand pedestal. Instead of providing an absolute 'God's-eye view' of the universe, space and time are entirely relative, dependent on the physical situations and interactions between the observer and the observed.

Once again, it's worth emphasizing that, by 'observer' we don't necessarily mean a human observer, though it's always worth bearing in mind that, in the final analysis, we're seeking a description of the physical world that we can comprehend with our distinctly human form of intelligence.

Of course, the observer plays a fundamentally important role in quantum theory, too. Suppose we are interested in making some measurements on a certain kind of quantum system. It doesn't really matter what this is. We've learned that applying the 'rules' of quantum theory means that we must begin by summarizing what we know about the physics of this system in the form of a wavefunction. This may have several component parts corresponding to different possible quantum states that the system can possess. Typically, the wavefunction will take the form of a superposition, which is just a fancy way of saying that we add up the contributions from all of the different states that are possible—let's say a certain amount of 'this' and a different amount of 'that'.

What we do next depends on what kind of measurement we propose to make. Let's imagine that our device is designed to reveal quantum states that are 'aligned' with some axis or field—up or down. Again, it doesn't matter precisely what this alignment is or how we interpret it: what matters is that the quantum states to be measured are determined by the quantum system and the nature of our device.

To make a prediction, we now need to use a few more of the rules of quantum theory to convert the wavefunction expressed in terms of a superposition of 'this' and 'that' into a superposition of 'up' and 'down'. This is called *changing the basis* of the wavefunction. We're perfectly at liberty to do this. Such freedom arises because we can build up a waveform from any set of suitable components, provided we follow the rules and add the components together in the right proportions.[6]

We suppose that, as a result of changing the basis, we get a wavefunction containing a bit of 'up' and a bit more 'down'. When we perform a measurement we can then expect to get 'up' with a probability given by the square of 'a bit', and we can expect to get 'down' with a probability given by the square of 'a bit more'.[7] If we've applied the rules correctly, these probabilities will sum to 100 per cent.

We now perform a measurement, the wavefunction 'collapses', and the system reveals itself to be in the 'down' state. We repeat the measurement with a series of identically prepared systems, and we record the sequence 'down', 'up', 'up', 'down', 'down', 'up', 'down', 'down', 'down'.... We find that we get 'up' with a frequency consistent with the quantum probability for this state, and the same for 'down'. All is well with the (quantum) world.

But then we get a little uncomfortable with all this because we've learned from the work of Bell and Leggett and the many extraordinary experiments that have been performed to test the predictions of quantum theory that it really makes no sense to ascribe any properties to the actual state of this system *prior* to measurement. We could assume that the quantum system is already in either an 'up' state or a 'down' state (perhaps due to the operation of some hidden variable which somehow tells it what kind of measurement we're going to perform), and the measurement simply tells us which it is. But we also know by now that if we assume this, then there are circumstances in which our predictions will be quite incorrect.

We're also uncomfortable because we have absolutely no physical explanation for the 'collapse' of the wavefunction—we just *assume* that we start out with a superposition and, somehow during the act of measurement, for each individual system this magically transforms from two potential outcomes—'up' *and* 'down'—to one actual outcome—either 'up' *or* 'down'.

Rovelli argues that our discomfort arises because we're looking at this in the wrong way, and it's much better if we stick with what we actually *know*. Our tendency is to think of these quantum states as *physical states of the system*. This is a logic that Rovelli suggests originates with Schrödinger's introduction of the wavefunction in the 'wave mechanics' of the hydrogen atom. In a bid to restore some sense of 'visualizability' in the physics of the atom that had been lost in Heisenberg's equivalent formulation (called matrix mechanics), Schrödinger was keen to establish the wavefunction as a literal, physically meaningful description:[8]

> The confusion got into full shape in the influential second paper...where Schrödinger stressed the analogy with optics: the trajectory of a particle is like the trajectory of a light ray: an approximation for the behaviour of an underlying real physical wave in physical space. That is, the [wave]function is the 'actual stuff', like the electromagnetic field is the 'actual stuff' underlying the nature of light rays.

Although Schrödinger's literal wave interpretation didn't survive, the notion that the wavefunction is loaded with physical meaning was dragged into the interpretation of the admittedly more obscure concept—the 'state vector'—that replaced it.*

But what we're really dealing with here is nothing more than the *relation* between the quantum system and the instrument we use to prepare it, and the subsequent *relation* between the prepared system and the measuring device.

We apply what we've learned from previous experiments and experience to summarize the wavefunction of the prepared system in terms of a basis which has components 'this' and

---

* To avoid confusion, I'll continue to use the term wavefunction here.

'that'. Our mistake, according to Rovelli, is to presume that such a wavefunction provides a *physical* description. It doesn't. It simply summarizes information that, if used properly within the rules, allows us to make accurate predictions. This information is derived from what we know about the system and the instrument used to prepare it. To make use of this information we need to translate it into a form that will apply to the specific measurement we want to make. In other words, we change to a basis determined by the relation between the prepared system and the measuring device. Now we're in a position to predict the probabilities of the different possible measurement outcomes.

In essence, Rovelli is telling us that we're mistaken in thinking that the wavefunction is 'real', at least in the sense that it represents the underlying physical properties of the system. 'The [wavefunction] that we associate with a system…is therefore, first of all, just a *coding* of the outcome of these previous interactions with [the system]'.[9] The structure of quantum theory—with all its wavefunctions and its rules—then serves as an instrument which enables us to predict future events on the basis of past events.*

This is entirely consistent with Bohr's position, summarized in a famous quote:[10]

> There is no quantum world. There is only an abstract quantum physical description. It is wrong to think that the task of physics is to find out how nature is. Physics concerns what we can say about nature.

And, for good measure, let's throw in another famous remark, this time from Heisenberg:[11]

---

* A moment's reflection will allow you to conclude that this kind of instrumentalist logic is not confined to quantum theory. By the same token, theories of classical physics can be interpreted in much the same way.

This again emphasises a subjective element in the description of atomic events, since the measuring device has been constructed by the observer, and we have to remember that what we observe is not nature in itself but nature exposed to our method of questioning.

Let's push this a little further. My reading of Rovelli's relational interpretation of quantum mechanics insists that reality is defined only in the context of the interaction between the system and the device that is 'making the observation'. Like everything in the universe, this device is also made of quantum objects. To another device (another observer), the 'measurement' of 'up' versus 'down' isn't a measurement at all. If we suppose that the first device has a pointer which points to the left if the system is in the 'up' state, and to the right if it is in the 'down' state, then another observer has no choice but to conclude that the original quantum system has now become *entangled* with the first device. Before looking to see which way the pointer goes, a correct summary of the information available takes the form of another superposition, this time of states formed from the combination of 'up' multiplied by left and 'down' multiplied by right.

We can go on like this forever, it seems, and this was precisely the point that Schrödinger was making with his famous cat paradox. But Rovelli is quite relaxed. If the device is rigged such that pointing left kills the cat, then the correct summary of the available information takes the form of a superposition of the product states of 'up' × left × dead and 'down' × right × alive. To discover the state of the cat we must introduce yet another device (me, or you) capable of lifting the lid of the box and looking.

Our instinct is to insist that, surely, the cat must already be either dead or alive *before* we lift the lid. Rovelli shrugs his shoulders. For sure, we can *speculate* about the physical state of the cat before the 'act of measurement' but we cannot escape a simple truth: we cannot *know* the state of the cat until we establish a relationship with it, by lifting the lid and looking.

Our mistake is to think that the superposition represents the cat's physical state—that the poor cat exists in some peculiar purgatory—rather than simply representing a summary of what we actually know about the situation. Lifting the lid doesn't 'collapse' the wavefunction, dragging the cat from purgatory into a state of deadness or aliveness. The only thing that changes when we lift the lid is our knowledge, and 'This change is unproblematic, for the same reason for which my information about China changes discontinuously any time I read an article about China in the newspaper'.[12]

This logic can be extended without much difficulty to an EPR-type situation involving entangled quantum states. Suppose our quantum system consists of two 'particles' which, by some law of conservation, are constrained such that if one (particle A) is 'this' then the other (particle B) must be 'that', and vice versa. As before, our measuring device requires us to translate these contributions to a basis of 'up' and 'down', and the physics constrains the outcomes such that if A is measured to be 'up', then B must be 'down', and vice versa.

The two particles separate and propagate some long distance apart such that they are no longer in causal contact (no physical influence travelling at the speed of light can pass from one to the other in the time available). In one laboratory, Alice observes that particle A is 'up'. Because she knows how the original quantum system was prepared, she can *speculate* that particle B must be 'down', but at the instant that particle A is observed she obviously cannot *know* the state of B. Likewise, Bob observes that particle B is 'down' but can only speculate that particle A must therefore be 'up'. For this situation to change a *further interaction* is required which could involve Alice and Bob communicating with each other to share their results. Or perhaps they both share their results with a third observer—Charles—who concludes from this that the states of the particles are indeed correlated—A is 'up' and B is 'down'.

All that changes through this sequence is the nature of the *information* available to Alice, Bob, and Charles. And, of course, the correlation between the results recorded by Alice and Bob, as established by Charles, simply reflects the information coded in the wavefunction when the original quantum system was prepared. All we've done is take information about a past event and used it to predict the outcome of a series of subsequent events.

This all seems perfectly reasonable and might even provoke something of a sense of relief. By accepting that quantum theory deals only with coded information that allows us to predict the future based on the past, we recover 'locality'—there is no 'spooky action-at-a-distance'—and we eliminate the need to invoke a 'collapse' of the wavefunction. We avoid the need to explain the transition from a microscopic quantum scale, with all its curious superpositions, to a macroscopic classical scale of cats and human observers. There simply is no transition. All we do is track the information coded in the wavefunction through the sequence of interactions required to establish our knowledge of the state of the system, whatever it is.

But Rovelli's relational interpretation requires us to make a significant trade-off, and there is a heavy price to be paid. To gain these advantages we must relax our grip on reality itself. We must accept that asking what a quantum system such as an electron 'really is', or even what physical state it is in at a given moment free of any interaction, is an exercise in pointless metaphysics. We must instead deal only with what we can know, and what we know can only come through interactions and experience.

This is a firmly empiricist or 'anti-realist' interpretation of the wavefunction. Not anti-realist in the sense that Rovelli is denying the existence of reality per se. In fact, a detailed, three-way e-mail exchange over several days among Rovelli, Smolin, and myself established quite clearly that Rovelli is a realist—he believes in the existence of an independent reality in which 'measurement' is no different in principle from any other kind of physical inter-action and there is nothing particularly special about interactions that create 'knowledge'. He is rather arguing that we should inter-pret the concepts we use routinely in quantum theory—and especially the wavefunction—not as literal expressions of real physical things but as convenient ways of communicating infor-mation about them, and so connecting past and future.[13]

There are precedents for this kind of logic. In the Critique of Pure Reason, first published in 1781, the great German philosopher Immanuel Kant distinguished what he called noumena, the meta-physical objects of reality or 'things-in-themselves' that can be conceived only in our minds, and empirical phenomena, the 'things-as-they-appear' in our perception and experience.* Kant claimed that it makes no sense to deny the very existence of things-in-themselves, as there must be some things that cause appear-ances in the form of sensory perceptions (there can be no appearances without anything that appears). But whilst the things-in-themselves must exist, we can in principle gain absolutely no knowledge of them.

---

* This doesn't mean that noumena are merely figments of a fertile imagination. I can imagine all kinds of things—such as unicorns—but these obviously have no place in our empirical reality. In the context of the present discussion noumena are things like electrons. We can point to lots of ways in which what we call 'electrons' manifest themselves in our empirical reality. These are electrons-as-they-appear. But an electron-in-itself without any kind of interaction through which it can make itself manifest exists, by definition, only in our minds.

For Einstein, the purpose of the EPR experiment was to lay bare the incompleteness of quantum theory. EPR opened their paper by providing a 'working definition' of physical reality:[14]

> If, without in any way disturbing a system, we can predict with certainty (i.e. with a probability equal to unity) the value of a physical quantity, then there exists an element of physical reality corresponding to this physical quantity.

This is, of course, a philosophically loaded statement.* According to Rovelli's relational interpretation of the wavefunction, no amount of certainty in the prediction of a physical quantity (a thing-in-itself) alters the simple fact that it *cannot* become part of empirical reality until it has undergone an interaction of some kind, allowing it to acquire properties *in relation* to that interaction (and so becoming a thing-as-it-appears). In fact, on this logic, 'without in any way disturbing a system' and 'element of physical [empirical] reality' are mutually exclusive. To become an element of empirical reality the system *must* be exposed to some kind of interaction.

I've been thinking and writing about quantum mechanics for more than 25 years, and I can tell you that the relational interpretation is the clearest articulation of the original Copenhagen position I've ever come across. It is certainly much clearer than many of Bohr's own pronouncements on the subject, which tended to rely on an obscure division between the microscopic quantum world and the classical world in which we make our observations

---

* And in fairness to Einstein we should acknowledge that the EPR paper was written largely by Boris Podolsky with—it seems—little input from Einstein himself. Einstein didn't agree with this definition as he knew it wasn't really needed to make the incompleteness argument and represented a weakness that would be easy for opponents to attack.

and perform measurements, or what John Bell would sometimes refer to as the 'shifty split'.[15]

But it still leaves many theorists—and most experimentalists—feeling very uneasy. Its logic is pretty irresistible, but it has us scrambling to find contentment with a formulation that is all about passing information about physical systems along a chain of interactions. We have to be satisfied with a structure that must remain stubbornly silent in the face of our—seemingly perfectly reasonable—questions. *Just how does nature actually do that?* Like emergency services personnel at the scene of a tragic accident, the interpretation advises us to 'move along: there is nothing to see here'. Some have argued that the Copenhagen interpretation tells us that such questions are quite pointless and we should instead just 'shut up and calculate'.[16] Rovelli echoes this sentiment:[17]

> I think that we should not try to keep asking what amounts to this same question over and over again: trying to fill-in the sparse ontology of Nature with our classical intuition about continuity.

Let's be quite clear that nobody here is going to argue with Kant. Nobody is saying that we expect ever to gain knowledge of quantum objects-in-themselves. But through their interactions these quantum objects project or impress themselves into our empirical reality of experience. Few will claim that we have completely exhausted our potential for gaining new knowledge of quantum objects-as-they-appear. We have not yet come to the end of science.

The relational interpretation says that the concepts and structure of quantum theory should not be taken literally as representing the 'real' states of quantum systems in the absence of interaction. This might be true, but the concepts and the structure are still telling us *something*. And it is in the nature of the game that scientists play with reality that, often by taking this 'something' at face value, we find that there is yet more to learn.

But let's be under no illusions. *The choice we face is a philosophical one.* It is no less than an appeal to faith, as Einstein once admitted: 'I have no better expression than the term "religious" for this trust in the rational character of reality and in its being accessible, to some extent, to human reason.'[18]

We can accept an anti-realist interpretation of the wavefunction, or we can adopt a more realist perspective, by investing the concepts and structure of quantum theory with a more literal, physical meaning. The price we pay for this realism is to be drawn irresistibly into what I've called the great 'game of theories'.* If empty empiricism is the rock shoal of Scylla, and Charybdis is a whirlpool of wild, unconstrained metaphysical nonsense about the nature of reality, then playing the game of theories is all about navigating the passage of the Strait of Messina in between. This means that speculating about the nature of a reality of things-in-themselves cannot be avoided. In doing this we might be led astray (as I believe proponents of interpretations based on many worlds and consciousness have been led astray). But I guess we must accept that it goes against the grain of human nature not to *try*.

Perhaps, just as Odysseus reasoned, it is better to risk a few crew members by sailing too close to Scylla than risk losing the whole ship in the whirlpool of Charybdis. If we are to learn something new, our speculations *must* connect with the empirical facts of the things-as-they-appear. After all, this is what it means for something to be scientific.

Taking the concepts and structure of quantum theory at face value means returning to the notion that the wavefunction *does*

---

* With acknowledgements to George R. R. Martin.

tell us something about the underlying physics and obliges us to puzzle over quantum probability, the nature of quantum measurement, non-locality, and the challenge of delimiting the quantum world of superpositions from the classical world of our direct experience. If we're prepared to do this, then I think there is no alternative but to accept that quantum theory is manifestly incomplete. The question then becomes: how could or should it be completed?

Einstein, Podolsky, and Rosen concluded their seminal paper with the following remark: 'we left open the question of whether or not such a [complete] description exists. We believe, however, that such a theory is possible.'[19] Einstein himself had toyed with hidden variable theories in 1927, only to reject the approach as 'too cheap'.[20] It seems likely that he thought that a solution could be found only in an elusive grand unified theory. In other words, that it might be possible to resolve the foundational problems of quantum theory by taking gravity properly into account.

But how, exactly? Penrose has long argued that a complete theory would necessarily include a physical explanation for the transition from many potential measurement outcomes to just one actual outcome. He calls it 'objective reduction'.* In *The Emperor's New Mind*, he argued that a correct quantum theory of gravity might hold the secret, that the trigger for the collapse might be a gravitational effect, caused by the curvature of spacetime:[21]

My own point of view is that as soon as a 'significant' amount of space-time curvature is introduced, the rules of quantum linear

---

* The 'collapse of the wavefunction' is also sometimes referred to as the 'reduction of the state vector'. The terminology is different, but the meaning (and so the nature of the problem) is the same.

superposition must fail. It is here that the complex-amplitude superpositions of potentially alternative states become replaced by probability-weighted actual alternatives—and one of the alternatives *actually* takes place.

Twenty-six years later, with the publication of *Fashion, Faith and Fantasy*, his views haven't changed but the logic of the proposed mechanism has been refined. He now suggests that it is the *difference* in the extent of gravitational interaction between the components of a superposition that causes it to unravel, 'collapsing' it into one or other of the possible outcomes. In this sense, the gravitational field provides a kind of 'frictional' resistance which leads to *decoherence*, a concept first introduced in quantum physics in 1970 by German theorist Dieter Zeh.[22] The superposition melts away as its components stop marching in tandem.

All mass-energy generates a gravitational field, so anything approaching a classical-size measuring device will generate more than enough interaction to ensure collapse. And, indeed, although entangled superpositions can be created in the laboratory and even transmitted over very large distances,* they are delicate and all-too-easily destroyed. The outcome of a measurement is then decided at a very early stage in the chain of interactions, long before it reaches the cat, or Gwyneth Paltrow. Penrose is hopeful that the timescale of gravitationally induced collapse might be such that it will soon be possible experimentally to catch a quantum system in the act.

This still leaves us with some explaining to do. Whilst it's good to get some sense for how the infamous collapse might make itself manifest, we're left to ponder exactly how the actual

---

* In June 2017, Chinese scientists successfully transmitted entangled photons from a communications satellite 500 kilometres above the Earth to two base stations 1200 kilometres apart.

measurement outcome is 'chosen' from among the various possibilities. John Bell, for one, was relatively unimpressed: 'The idea that elimination of coherence, in one way or another, implies the replacement of "and" by "or" is a very common one among solvers of the "measurement problem". It has always puzzled me.'[23]

Sorting out the collapse problem is one thing, but if we're intent on finding a deterministic solution for the inherent non-locality of quantum mechanics, then that's quite another. Elaborate experiments have established correlations between the quantum states of entangled particles over distances much, much greater than can be reached by signals travelling at the speed of light. If the wavefunction collapses and particle A is 'up', how does particle B adopt a 'down' state when it cannot possibly be in any way influenced by what has happened to A?

It seems we have no choice but to invoke some form of hidden variables. But we know from all the experimental tests of Bell's and of Leggett's inequalities that, if they exist, then these *must* be non-local, and so just as 'spooky' as quantum theory itself (and 'too cheap', into the bargain).

But wait.

In a quantum theory of gravity in which spacetime is emergent, just what is 'locality' anyway? Of course, we think of something as being local if it is 'here', confined to a small region of space. In the context of LQG, such a region might be defined in terms of a group of adjacent points on the dynamic lattice emerging from a spinfoam. But beneath this lattice sits an evolving spin network consisting of nodes and links. And, as we learned in the previous chapter, it is possible to model the propagation and interaction of fermions over such networks. So, what does 'locality' mean in this context?

It seems that we now have two notions of locality, as Markopoulou and Smolin explain:[24]

> There is a notion of locality in the [spin network] graph of the fundamental theory: two nodes are nearby if they are connected in the graph. A separate notion of locality holds in the embedding of the graph used in the low energy limit [in which the nodes become coordinate positions]. Two particles, represented by the embedding of two nodes in the graph, are nearby if they are close in the metric of the embedding space.

In LQG the nodes (locations of the quanta of volume) map to locations or lattice points in the emergent space (through a process of 'embedding' in the above quote). If the mapping is 'orderly', nodes that are adjacent and linked in the spin network translate to locations that are adjacent in the space that emerges— we get an orderly lattice as shown in Figure 24a. A particle propagates by 'hopping' from one location to the next, so in the spacetime that emerges it takes some time for the particle to cover large distances, explaining why the speed of light is finite.

But the mapping doesn't have to be orderly. American theorist Murray Gell-Mann used to say that in quantum theory 'everything not forbidden is compulsory'.[25] And, indeed, links in adjacent nodes can translate in LQG into links between non-adjacent spacetime locations, as shown in Figure 24b. Here, 'non-adjacent' could mean halfway across the universe. Of course, there can't be too many of these non-local links, or we would be very (perhaps painfully) aware of them.* 'Still, nodes connected non-locally to somewhere across the universe would be fairly common', suggests Smolin, 'there would be, on average, more than one per cubic nanometre of space.'[26]

Recall from the previous chapter that in their efforts to model fermions, Rovelli and his colleagues observed that the fermion

---

* And too many non-local links would make it impossible for a recognizably 'continuous' space to emerge.

(a)

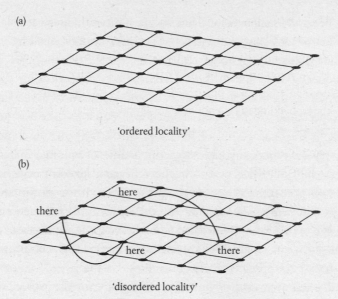

'ordered locality'

(b)

'disordered locality'

Figure 24. In LQG the nodes of a spin network map to locations or lattice points in the emergent space. An orderly mapping results in an orderly lattice (a). But the mapping doesn't have to be orderly: links in adjacent nodes can translate to links between non-adjacent spacetime locations (b). This is 'disordered locality'.

loop is sometimes non-local: it can disappear from one location to reappear at some distant location. Smolin and Markopoulou have called this 'disordered locality'.

Even if it really does exist, it's very unlikely that we'll ever be able to detect the effects of disordered locality in the laboratory. Although some simple probability estimates suggest that there may be as many non-local links as there are baryons in the universe, any such link in the space inside a laboratory here on Earth will most likely randomly connect with another point well outside our own galaxy.[27] There would be no way of telling.

But could such non-local links be the ultimate hidden variables? When we say that two entangled particles are described by the same wavefunction do we really mean that the two particles are connected by a link in the underlying spin network? One that persists as the particles propagate over the spinfoam and separate in spacetime? Could it be that when we see that particle A is 'up', the link acts as a kind of backdoor through spacetime which somehow causes particle B to adopt a 'down' state? We sit back scratching our heads wondering how the particles communicated *through* spacetime at speeds faster than light, when all the time they stayed in causal contact with each other *beneath* spacetime.

We've thus far been inclined to think that a quantum theory of gravity is one in which quantum properties and behaviour have primacy and spacetime and general relativity emerge as secondary, low-energy approximations or limits. In other words, we assume that a quantum theory of gravity is first and foremost a quantum theory. But if we take non-locality as a fundamental property or behaviour of quantum objects, then from the above discussion we can see that there is a sense in which quantum theory itself also emerges as secondary.

Perhaps LQG is hinting at a rather radical insight. Maybe there is one theory based just on abstract networks of interactions from which *everything* is emergent—the quantum properties and behaviour of matter and radiation *and* spacetime, the gravitational field. Markopoulou and Smolin showed in 2003 that it is indeed possible to deduce non-relativistic quantum theory and spacetime from the same low-energy limit of a background-independent model, and Smolin continues to pursue this idea.

14

NOT WITH A BANG
The 'Big Bounce', Superinflation, and Spinfoam Cosmology

After Lee Smolin's departure from Penn State and Carlo Rovelli's return to Europe, as I mentioned in Chapter 11 Ashtekar's interests turned increasingly towards black hole physics and cosmology. These interests were principally founded in classical theory, but he was inevitably drawn to the challenge of applying the new description of space and time offered by LQG.

Recall that Einstein found that he could apply his general theory of relativity to the entire universe, so risking confinement in a madhouse. According to the standard model of inflationary Big Bang cosmology, the universe 'began' in a hot big bang, an explosion into existence of space, time, and energy, which we believe occurred about 13.8 billion years ago. Its proponents argue that the model resolves the flatness and horizon problems by invoking a short burst of exponential inflation, between about 10^{-35} and 10^{-32} seconds after the Big Bang. The idea is that this rapid expansion imprinted quantum fluctuations that were occurring in the inflaton field onto the larger structure, leaving telltale traces, among them the bloody thumbprint of tiny temperature variations that we can detect in the cosmic background radiation.

As we have already seen, this doesn't quite give us everything we need. To complete the model we need to add two rather important but very mysterious ingredients, which together account for no less than 95 per cent of the total mass-energy of the universe: dark matter and dark energy. The latter manifests itself as a small, positive cosmological constant, responsible for the gentle acceleration in the expansion of the universe that we can detect today.

On the one hand, the model is extraordinarily pleasing. With only six variable parameters we can reconcile a number of disparate cosmological observations and bring them together into a marvellous unity. But on the other hand, the model is extraordinarily frustrating. We can't explain most of it. And we can't apply our theories to the very 'beginning', as they break down completely at the moment of the Big Bang itself, the point where all of known physics comes to an abrupt end.

This happens because general relativity is a classical (non-quantum) theory, and so admits the possibility of things that are infinitely large or infinitesimally small. In the context of the Big Bang, the infinite is manifested as a singularity, the point at which the mass-energy density of the universe is infinitely large and the entire universe is compressed into the infinitesimally small. A singularity is really a bit of an embarrassment.

It seemed that there was no easy way to spare ourselves this embarrassment. Some theorists argued that singularities arise because of the highly symmetric nature of the spacetime metrics used to provide generic solutions to Einstein's gravitational field equations, and that they would disappear in other, much less symmetric, solutions. But, in 1970, Penrose and the young Stephen Hawking at Cambridge University in England devised a series of *singularity theorems* which established that, in *any* attempt to apply general relativity at extremes of mass-energy density and spacetime curvature, a singularity is *inevitable*.

The prediction of a Big Bang singularity means that we can't even *ask* our perfectly legitimate and rather burning questions about the origin of everything, let alone answer them.*

And this is really only the beginning of our troubles. Cosmic inflation appears to resolve the flatness and horizon problems but still leaves us scratching our heads about the initial conditions, and the duration of the inflationary period required to produce the universe we observe. It resolves one fine-tuning problem only to give us another. Some theorists now argue that the most recent Planck satellite data on the cosmic background radiation actually strongly *disfavour* the simplest, 'slow roll' inflation scenarios.[1] And, of course, current theories offer no clues as to why the cosmological constant has the value that it does.

Let's take these problems one at a time.

As LQG emerged as a satisfyingly consistent theory of quantum gravity in the late 1990s, Ashtekar was aware of its implications for our understanding of the very earliest moments in the evolution of the universe, and particularly the problem posed by the Big Bang singularity. As he explained, '*the big bang is a prediction of general relativity precisely in a domain where it is inapplicable*'.[2]

Hawking had understood that general relativity is incomplete and, in *A Brief History of Time*, he wrote:[3]

> However, general relativity claims to be only a partial theory, so what the singularity theorems really show is that there must have been a time in the very early universe when the universe was so small, that one could no longer ignore the small-scale

* It's worth noting that the current quantum field theories that collectively form the standard model of particle physics are adequate to about a trillionth of a second after the Big Bang, at which time interactions with the Higgs field are believed to have triggered symmetry-breaking and a separation of the weak and electromagnetic forces. I've always thought that being able to extrapolate back to a trillionth of a second after the notional beginning of everything is pretty impressive, but I guess some folks are never satisfied...

effects of the other great partial theory of the twentieth century, quantum mechanics.

The early versions of quantum geometrodynamics and the Wheeler–DeWitt equation didn't provide useful alternative ways of looking at the problem. Hawking believed that the singularities should disappear in a fully fledged quantum theory of gravity, but in 1988 he was unsure how such a theory could be constructed.

There the problem sat, for a further 11 years. Then, Martin Bojowald, a young Ph.D. student at the University of Aachen in North Rhine-Westphalia, published a couple of papers that caused something of a stir. These were rather technical papers, but they demonstrated the possibility of developing a quantum cosmology based on LQG.

This was extraordinarily exciting. At the back of everybody's mind was a simple question: Would the discrete quanta of space demanded by LQG eliminate the Big Bang singularity altogether? If it did, then the resulting loop quantum cosmology (LQC) might continue to be applicable all the way back to 'time zero'.

Bojowald completed his Ph.D. in 2000, and Ashtekar and Smolin arranged to bring him to Penn State as a postdoc. Bojowald set to his task. The question he confronted was easily stated—'What happens to the classical cosmological [Big Bang] singularity in quantum geometry?'[4] The answer was not quite so straightforward. It involved a few assumptions and mathematical manipulations but he was able to satisfy himself of the answer: 'For small volume, quantum geometry leads to new effects which are responsible for the *removal* of the classical singularity.'[5]

Ashtekar was impressed. 'This was a very big deal,' Rovelli claimed. 'Everyone had hoped that once we learned to treat the quantum universe correctly, the big bang singularity would disappear. But it had never happened before.'[6]

LQC had arrived, but it wasn't all plain sailing.

We saw in Chapter 5 that it is possible to derive exact solutions to Einstein's gravitational field equations, each representing a different spacetime metric. In the context of cosmology, these solutions can be written in terms of something called the 'cosmic scale factor', a dimensionless quantity usually given the symbol a. We can think of the scale factor simply as the 'proper distance' (say, between two galaxies) at some time t, divided by this distance at some earlier fixed time, t_0. In an expanding universe, this distance *increases* with time (look back at Figure 13 in Chapter 5). Galaxies are carried further and further apart as the spacetime they're sitting in expands, so the scale factor a is time-dependent.

In a universe which expands forever at a constant rate, the scale factor increases linearly with time. This is the basis for Hubble's law: the velocity at which a galaxy is measured to be receding from our vantage point on Earth is directly proportional to its distance from us.[7]

The constant of proportionality in Hubble's law is called the Hubble parameter, H, which is just the ratio of the rate of change of a with time (usually given the symbol $\dot{a}$), divided by a. We can think of $\dot{a}$ as a *velocity*, of expansion or contraction. In a universe expanding at a constant rate, $\dot{a}$ is fixed and greater than zero. In a universe which is contracting at a constant rate, $\dot{a}$ is fixed and less than zero.

But, as we've seen, although our universe is expanding, the rate of expansion is *not* constant—the velocity $\dot{a}$ is itself increasing with time, or the expansion is *accelerating*. This means that the Hubble parameter is not constant—it varies over time. The Hubble *constant*, H_0, is then then value that the Hubble parameter possesses today.*

---

* And thanks to Alejandro Corichi for clearing up some confusion about this on my part.

It was soon realized that the early LQC solutions developed by Bojowald had some rather unphysical implications for the Hubble parameter. Nevertheless, the interest grew as LQG techniques were applied more systematically in a cosmological framework. Ashtekar worked with Bojowald, Jerzy Lewandowski, Parampreet Singh, Tomasz Pawlowski, and Alejandro Corichi, among others. More subtle and sophisticated models were derived.

Singh and Pawlowski developed a complete mathematical and conceptual structure and used computer simulations to provide numerical solutions. These simulations could be run forwards or backwards in time. As the theorists were interested in what might have been happening in the universe around the time of the purported singularity, they ran the simulation in reverse, watching as spacetime contracted, heading for a 'big crunch'. Ashtekar was completely taken aback by what they saw.

The backwards-running universe behaved pretty much as expected but, as it approached extremely high densities, instead of collapsing to a singularity it started to expand again.

The universe *bounced*.

But there were problems with this model, too. Having glimpsed a potentially extraordinary solution to the singularity problem, Ashtekar was initially rather downhearted. But, with Singh and Pawlowski, he continued to plug away. More detailed examination showed that the model had three unphysical features. Ashtekar later noted that:[8]

> On thinking more deeply, I realized that we were implicitly applying the LQG quantum geometry ideas to a background, reference metric rather than to the physical metric. When we redid the construction by applying them to the physical metric, all three problems were solved in a single stroke.

The reconstruction involved a change of perspective, from area to volume. In this way LQC went through four stages of evolution over the course of about six years before the theorists were satisfied that the structure is physically sensible. The good news was that the conclusions regarding the elimination of the Big Bang singularity held up. At least in these LQC model representations, the Big Bang is replaced by a 'big bounce' (Figure 25).

The origin and nature of the bounce are relatively easy to grasp in a so-called 'semi-classical' approximation to the full quantum version of LQC, in which the effects of quantum geometry manifest themselves as a 'correction' to classical general relativity. As Corichi explained: 'This model gave us a lot of power, since we could now ask so many questions and have exact answers.'[9] The resulting LQC solutions look very similar to those of general

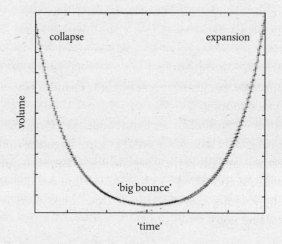

Figure 25. In loop quantum cosmology the existence of fundamental quanta of area and volume mean that the Big Bang singularity is avoided. As the universe collapses, its volume shrinks as shown on the left of this diagram. But when it hits a critical density (centre), the universe 'bounces' and starts to expand once more.

relativity based on familiar spacetime metrics.* A single additional quantum correction term appears. This term is $-\rho/\rho_c$, where ρ is the density of mass-energy in the universe and ρ_c is the 'critical' density at which the bounce occurs. The negative sign is important. In a contracting universe, with $\dot{a}$ less than zero, when the density of mass-energy reaches the critical density, ρ/ρ_c becomes equal to 1, terms in the equations cancel out, and $\dot{a}$ becomes equal to zero. The universe stops contracting.

The critical density ρ_c is closely related to the Planck scale density (which is about 5×10^{96} kilograms per cubic metre).[10] For sure, this is enormous—the density of matter at the centre of the Sun is estimated to be just 1.6×10^5 kilograms per cubic metre, and to reach the Planck density we would need to compress a trillion Suns to the size of a single proton. Yes, it's big, *but it's not infinite*. The very existence of quanta of volume prevents the universe from collapsing to densities greater than the Planck density. Quantum mechanics does not allow the singularity to be reached.

To understand why, it's helpful to reach for a few lessons from subatomic physics. A hydrogen atom consists of a single positively charged proton (the nucleus) and a single negatively charged electron. As Schrödinger himself discovered in 1925, in a quantum-mechanical description the lowest-energy state or 'orbital' of the electron in a hydrogen atom is represented by a spherical wavefunction with the proton sitting at its centre. We can use this wavefunction to tell us that the most likely place to 'find' the electron inside the atom is at a distance of 5.29×10^{-11} metres from the proton, a distance called the Bohr radius, as shown in Figure 26a. Note

* Interestingly, the exactly soluble version of LQC approximates the Wheeler–DeWitt equation for large volumes, but the latter cannot be recovered from the former by extrapolating the quanta of area and volume to zero (equivalent to extrapolating Planck's constant h to zero). This semi-classical version of LQC is still a fundamentally *discrete* theory.

(a)

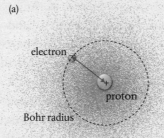

(b)

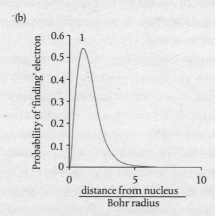

Figure 26. (a) In the old planetary model of the atom, the electron was presumed to orbit the proton at a distance called the Bohr radius. But quantum mechanics replaces the orbiting electron with an electron wavefunction. The electron can now be 'found' anywhere within this wavefunction, but it has the highest probability of being found at the distance predicted by the old planetary model (b).

the words 'most likely'. In quantum mechanics, the orbit of the electron is not fixed at a specific distance (in the same way that the Earth moves around the Sun with an orbital radius of about 150 million kilometres), but is instead 'spread' over a range of distances measured from the centre, reflecting the wave nature of the electron. This spread peaks at the Bohr radius—Figure 26b.

This might be puzzling for any number of reasons, but we know that electrical charges of opposite sign attract, and the force of attraction increases as charged objects get closer together. So why doesn't the negatively charged spherical wavefunction simply collapse down onto the positively charged proton? The result would not be a singularity, but it would be no less disastrous, at least for anything in this universe that depends on the existence of stable atoms.

It turns out that Heisenberg's uncertainty principle places a rigorous limit on the extent to which the electron wavefunction can be 'confined'. If we try to squeeze it down to smaller and smaller volumes around the proton—pinning it down to a more and more precise position—we simply drive up the uncertainty in its momentum. The wavefunction 'pushes back', exerting a kind of repulsive pressure which prevents it from being squeezed any further. Of course, the uncertainty principle depends on the size of Planck's constant h which, though very small, is not zero. And the quantum of area and volume in both LQG and LQC depend on the square and cube of the Planck length, respectively, which in turn depend on the size of h.

Ashtekar wrote:[11]

> The suggestion from LQC is that a new repulsive force associated with the quantum nature of geometry comes into play and is strong enough to counter the classical, gravitational attraction, irrespective of how large the mass is. It is this force that prevents the formation of singularities.

This kind of quantum repulsion is completely negligible when the curvature of spacetime is slight, say one per cent of the curvature associated with the Planck scale. In fact, for mass-energy densities ρ much smaller than ρ_c, the ratio $-\rho/\rho_c$ becomes vanishingly small and the semi-classical LQC solutions become identical to those of general relativity. But as the mass-energy density and spacetime curvature increase towards ρ_c, the resistance from the quantum nature of space itself starts to build up.

Of course, the approximations required to formulate this semi-classical version of LQC can be expected to break down as the density approaches the critical density. But in 2007, Ashtekar, Pawlowski, and Singh demonstrated that the same conclusions hold in simplified quantum versions of LQC as well.

So, according to LQC, our universe 'began' not with a bang, but with a bounce, which occurred about 13.8 billion years ago. Despite what some antitheists might want to claim about the creation of a universe 'from nothing',[12] it seems that the universe that we know today might have been created from the universe that existed *before* the bounce.

What then came before the Big Bang? The answer, based on what we can learn from LQC, is obvious: another universe, which contracted towards a big crunch but then bounced into an expansion phase. This begs an obvious next question: What (if anything) can we discover about the universe that went before?

In 2007, Bojowald's answer was 'Not very much.' He concluded that most of, though not all, the information about the previous universe was lost as it passed through the bounce. He called the phenomenon 'cosmic forgetfulness'.[13] No two universes either side of a bounce would ever be the same, not just in terms of their structure and appearance, but possibly also in terms of their physical laws and spectrum of particles. We can look all we want, but we'll never be able to learn anything about what came before the bounce.

But other LQC theorists did not share this view. In 2008, Corichi and Singh argued that, on the contrary, there are very strong bounds on the quantum fluctuations of the universe emerging from the bounce, such that 'the universe maintains an (almost) total recall'.[14] Further research published since 2008 supports this conclusion. To all intents and purposes, the universe that existed before the bounce is much the same as the universe we see today. For sure, stuff gets scrambled as it squeezes through the narrow neck defined by the Planck density, but there are grounds for thinking that the earlier universe might have possessed reasonably familiar laws and particles, which—who knows?—may get tweaked slightly with each bounce.

Bojowald holds firm to his original conclusion, arguing that quantum fluctuations in the distribution of *matter* (rather than spacetime) cause the universe to forget.[15] The question remains unresolved but, as far as I can tell, the consensus in the LQC theory community favours recall over forgetfulness.

By nailing down the details of a consistent LQC in this way, Ashtekar and his colleagues had opened up a lot of potentially exciting new possibilities relating to the physics of the early universe. In August 2007, Ashtekar became the Director of the multidisciplinary Institute for Gravitation and the Cosmos, the successor to the Center for Gravitational Physics and Geometry at Penn State. Today the institute has three centres: for fundamental theory, for theoretical and observational cosmology, and for particle and gravitational astrophysics.

For a time, it seemed as though the physics of the bounce was all that was required to provide the rapid burst of cosmic inflation believed to be necessary to resolve the flatness and horizon problems. This would have been a fantastic conclusion, as the

Λ-CDM model is obliged to *assume* that inflation happens—the inflaton field is added to the equations 'by hand' without an obvious natural origin or cause. Whilst this might seem quite artificial and arbitrary, we must remember that much the same trick was played with the Higgs field in the standard model of particle physics (and look how that turned out).

In LQC, the universe emerging from the bounce undergoes a very rapid burst of expansion, which the theorists called *superinflation*. Ashtekar noted that 'Superinflation never occurs in general relativity, whereas it is *compulsory* in loop quantum cosmology.'[16] In 2002, Bojowald had argued that this is all the inflation the universe needed, and that superinflation furthermore accounts for the residual expansion and the small cosmological constant we observe today.[17] In the more recent LQC models, superinflation occurs as the post-bounce universe starts to expand, and as the mass-energy density falls from ρ_c to $\frac{1}{2}\rho_c$.

But it was soon realized that superinflation could be no substitute for the slow roll inflationary scenario adopted in the standard Λ-CDM model. It simply doesn't last long enough to have the desired effects. LQC on its own can't explain the large-scale structure of our universe. This was rather disappointing, and it seemed that there was no alternative but to go with the flow and supplement LQC with an inflaton field with much the same properties assumed in the standard inflationary Big Bang model.

There was a bonus, however. Incorporating inflation arguably helps to resolve the fine-tuning associated with the flatness and horizon problems. But, as I mentioned earlier, the way it must be applied implies further fine-tuning of its own. The post-inflation structure of the universe is extremely sensitively dependent on the conditions that prevailed going into inflation. Given the expected nature of the quantum fluctuations in the inflaton field, the chances of getting just the right amount of inflation and exiting with a structure that will lead inexorably to the stars, galaxies, galaxy clusters, and voids familiar in our universe would appear to be extremely low. Estimates of the probability of getting to a universe just like ours vary considerably, but most put the chances at 1 in 10 to the power of a ridiculously large number.[18] Our post-inflation universe would appear to be so implausible that we need to accept an extraordinary degree of fine-tuning of its initial conditions if we are to explain why it exists at all.

We could just shrug our shoulders at this point and acknowledge that 'since we have only one universe, the issue of likelihood is irrelevant'.[19] But physicists get rather vexed by coincidences they can't explain, especially when they resemble some kind of cosmic conspiracy.

And here's the pay-off. Supplementing LQC with the same kind of inflaton field used in the standard Λ-CDM model turns what at first appears to be an extremely unlikely outcome into a *near-certainty*.*

The origin of the difference is subtle. In conventional general relativity with added inflation, the presence of the singularity at the Big Bang means that it is not possible to get a fix on the initial conditions that prevail before inflation sets in. Literally anything can happen. This leads to the situation where we have to invoke fine-tuning to get just the right amount of inflation consistent with our observations of the universe today. But in the LQC big bounce scenario, the point of origin and the initial conditions are *much* better defined. And the physics of the bounce and superinflation mean that the universe is 'funneled to conditions which virtually guarantee slow roll inflation with [sufficient duration].'[20]

In other words, the universe is pushed in the 'right' direction, along trajectories drawn to what is known as a dynamical

---

* With a probability greater than 99.999 per cent. See endnote 20.

attractor, and a sufficiently long period of slow-roll inflation kicks in when the density has fallen some eleven orders of magnitude below the critical density at the bounce. No matter what we start with at the moment of the bounce, the presence of the attractor ensures that we end up with more or less the universe we do, in fact, observe. Ashtekar and David Sloan found that even adjusting the mass-energy of the inflaton field—another fine-tuned parameter in the Λ-CDM model—makes little difference to the onset and duration of the inflationary period.

Yes, inflation still must be assumed, but the physics of the bounce and superinflation make assumptions about the initial conditions much less arbitrary.

One last point. Although LQC can offer us no clues regarding the origin of the small, positive cosmological constant, it can be formulated using the Friedmann–Lemaître–Robertson–Walker (FLRW) metric in just the same way as in the standard Big Bang model. The inclusion of Λ does not affect any of the conclusions concerning the singularity, superinflation, and the dynamics of slow-roll inflation.

I guess two out of three isn't so bad, at least for now.

This sounds all well and good, but perhaps now is the time for you to ask the question that's been nagging at the back of your mind. All this talk about bouncing universes, superinflation, and slow-roll inflation assumes that these are all events that unfold in time. But how are we meant to understand this in a theory of quantum gravity from which time *disappears* or is supposed to be emergent?

The problem of 'frozen time' has certainly not gone away. In the original Wheeler–DeWitt equation, the size of the scale factor itself was used to mark time, but its nonlinear behaviour in LQC means that this is not a practical option (our experience is that time ticks by in a uniform, linear fashion, and in LQC the scale factor doesn't). Ashtekar, Pawlowski, and Singh found that they needed to introduce a new massless field to act as a kind of internal cosmological clock, providing a dimension which functions as an emergent time against which moments in the evolution of the universe can be mapped. As Ashtekar and Singh later explained: 'Such an internal or emergent time is not essential to obtain a complete, self-contained theory. But its availability makes the *physical meaning* of the dynamics transparent and one can extract the phenomenological predictions more easily.'[21]

But, hold on. In Chapter 10, I explained that time can be recovered in LQG by using the spinfoam approach, in which spacetime emerges as a superposition of sums-over-histories of transitions between the nodes and links of the dynamic spin networks. Why not use this approach to develop a *spinfoam cosmology*, dispensing with the need for an arbitrary massless field to keep track of time?

This might seem like a rather obvious question, but by 2007 it simply hadn't occurred to those theorists closest to the spinfoam formalism to ask it. 'None of us was doing cosmology in Marseille at the time,' explains Rovelli. Then a young Italian undergraduate called Francesca Vidotto arrived at the Centre for Theoretical Physics on a one-year Erasmus scholarship.* She expressed her total astonishment that nobody was working on spinfoam cosmology. 'Two or three years later we were all doing quantum cosmology in Marseille.'[22]

Rovelli and Vidotto got to work, publishing their first papers on the cosmological implications of spinfoams in 2008 and 2009, and taking their first steps towards a formal spinfoam cosmology in collaboration with Eugenio Bianchi in 2010. Ashtekar

---

* The Erasmus Programme is a European Union exchange student programme that started in the late 1980s. It provides scholarships to support students in getting placements at foreign universities within the EU.

joined the party with a paper published independently in 2009, in collaboration with Miguel Campiglia and Adam Henderson.

Most of these papers are principally concerned with technical issues associated with the application of the spinfoam approach, and this is still very much a work in progress. The theory is not yet sufficiently mature to provide detailed insights on the physics of the universe close to the Big Bang. However, it does provide very strong hints that, once again, the classical Big Bang singularity is avoided. The existence of quanta of area places a restriction on the extent to which the contraction of the universe can be accelerated. Instead of collapsing to a singularity (infinite acceleration), it hits a buffer at a mass-energy density close to the Planck scale.[23]

There is more to come, and this is a space that should probably be watched. Closely.

To a very considerable degree, LQC gives us a picture of a rather *more elegant universe* than that described by the standard Big Bang model (or, for that matter, by any string cosmology). At least, LQC provides a more elegant description or explanation. It eliminates the Big Bang singularity and, although we still need to incorporate an inflationary scenario, the big bounce and subsequent burst of superinflation pushes the universe in the desired direction, in an entirely natural way without the need for elaborate fine-tuning of the initial conditions. But the question remains: What difference does this description make to the way we interpret the things we can observe today, or the things we might hope to observe in the not-so-distant future? Without an answer to this question then, elegant or not, we are still arguing on the basis of aesthetics.

There's some hope. The physics of the Planck scale is judged by many to be completely beyond the reach of any apparatus we might one day install in a terrestrial laboratory.[24] But the universe is no longer a playground only for theologians, philosophers, or theoretical physicists. As we've seen, it is also fast becoming the ultimate playground for experimental science.

Replacing the Big Bang with a bounce potentially opens a window on the Planck-scale physics that occurred shortly afterwards. This physics leaves its mark, changing the nature of the fluctuations imprinted on the cosmos by slow-roll inflation, and hence on the cosmic background radiation, when compared with the standard Big Bang model.

The changes are very subtle, however. Figure 27 shows the variation in the square of the temperature differences in the cosmic background radiation across the sky, as recorded by the

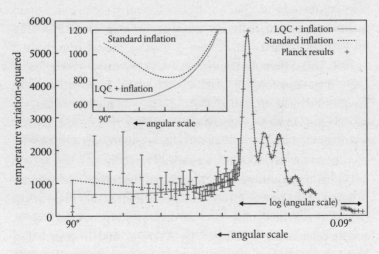

Figure 27. The variation in the squares of the temperature fluctuations with angular scale across the sky, as recorded by the Planck satellite and reported in 2013. The data are the points, with associated error bars. The solid curve is the prediction of LQC combined with slow-roll inflation, and the dashed curve is the prediction of the standard inflationary Big Bang model. The predictions diverge most noticeably at large angular scales, where the errors bars in the Planck results are largest. Nevertheless, the LQC/inflation model provides a better fit to the existing data.

Planck satellite, as a function of angular scale (which decreases from left to right).[25] This shows a pattern consisting of one large peak, followed by two smaller peaks of equal height, followed by a sequence of progressively damped oscillations as the angular scale decreases. This pattern—called a power spectrum—can be reproduced using simple hydrodynamic models which treat the universe as a fluid. The oscillations are then impressions left by *sound waves* that bounced back and forth across the universe at the time of recombination.*

The big bounce of LQC and the Big Bang of the standard model predict subtly different patterns. These differences are most noticeable at large angular scales, to the left in Figure 27. Here LQC combined with slow-roll inflation (shown as the solid curve) *suppresses* the extent of the temperature variation when compared with the standard inflationary Big Bang model (the dashed curve). This difference is more visible in the inset, which shows the calculations for large angular scales only. For small angular scales, both LQC and the standard Big Bang model predict the same results (note that the axis on the right in Figure 27 becomes logarithmic in angular scale, which serves to stretch out the picture on the left, allowing for closer inspection).

Unfortunately, these models predict differences which appear in a part of the power spectrum for which the standard errors in the most recent Planck satellite data are large. Nevertheless, recent calculations published by Ashtekar and Brajesh Gupt, building on earlier work by Ashtekar, Ivan Agullo, and William Nelson, suggest that the model based on LQC and slow-roll inflation predicts a power spectrum which provides a *better fit* to the existing Planck data than the standard Big Bang model.[26,27]

---

* I like to think that the universe was *singing*, although on a bad day *screaming* might be considered more appropriate.

Clearly, to be definitive we can no longer rely on our cosmic thumbprint. We need more sophisticated measurement: the cosmological equivalent of DNA fingerprinting.

But there's a little more. Inflationary scenarios leave telltale signatures in the form of so-called *primordial* gravitational waves. These are ripples in spacetime arising directly from inflation: they are not the same as the gravitational waves first detected by the LIGO observatory in 2015, which arise from the merger of black holes or neutron stars occurring much later in the universe's history.

Primordial gravitational waves are extremely elusive. They can potentially be observed in subtle, so-called *B-mode polarization* of the cosmic background radiation. The announcement, in March 2014, that such polarization effects had indeed been detected by the BICEP2* experiment generated considerable excitement. This soon evaporated when, in September that year, scientists working on the Planck satellite demonstrated that the BICEP2 observations could all be attributed to interference from cosmic dust particles.

There is no alternative but to look for primordial gravitational waves directly, rather than indirectly via subtle effects in the cosmic background radiation. Such a search requires a space-based observatory, and the good news is that following a successful pathfinder mission, in June 2017 the European Space Agency gave the go-ahead for the Laser Interferometer Space Antenna (LISA). This will involve placing a triangular array of three laser interferometers in orbit around the Sun, each spaced 2.5 million kilometres apart, trailing in the wake of Earth's orbit. Small differences in the distances measured by laser beams bounced from

---

* BICEP stands for Background Imaging of Cosmic Extragalactic Polarization. BICEP2 was a second-generation telescope based at the Amundsen–Scott South Pole Station in Antarctica.

one interferometer to the other will betray the passing of gravitational waves. LISA is expected to launch in 2034.

Of course, we anticipate that there are many different sources of gravitational waves, so, if they do indeed exist, the primordial waves will contribute to a kind of background 'noise'. If we are to put all the different theories about the primordial universe to the test, we will need to be able to analyse this noise in considerable detail. One advantage of LQC is that the spectrum of low-energy primordial gravitational waves is predicted to be distinctly different when compared with that of the standard Big Bang model, a feature that can be attributed directly to the bounce and superinflation.[28]

So, present observations of the cosmic background radiation can't yet give us a definitive answer, but the accuracy of these observations is improving all the time. And, in another 20 years or so, we may be able to observe the spectrum of primordial gravitational waves in detail (if they exist). 'The situation is still fluid,' says Rovelli. 'But those who, like myself, have spent their lives seeking the secrets of quantum space are following with close attention, anxiety and hope the continuous honing of our capacity to make observations, to measure and to calculate—and are awaiting the moment in which nature will tell us whether we are right or not.'[29]

15

BLACK HOLE ENTROPY, THE INFORMATION PARADOX, AND PLANCK STARS

Brian Greene compared counting the number of possible Calabi–Yau shapes that could be admitted in superstring theory with counting the number of grains of sand on 'every beach'.[1] He really wasn't kidding. In 2003 theorists Shamir Kachru, Renata Kallosh, Andrei Linde, and Sandip Trivedi figured out how they could stabilize superstring theory's hidden dimensions by threading lines of force through the holes in the Calabi–Yau space, so making the structure at least consistent with a universe a bit like our own, with a small, positive cosmological constant.[2]

But it turned out that there is a rather alarming number of ways this can be done. This is no longer just about the number of different possible Calabi–Yau spaces (the number of grains of sand). It is about the number of possible *theories* that can be deduced based on different ways of threading the lines of force through the holes. This threading determines the nature of the superstring vibrations that are possible and thus the physical constants, the laws of physics, and the spectrum of particles that will prevail. In other words, each possible theory corresponds to a different *type of universe*.

How many? There are ten different ways of threading the lines of force for a Calabi–Yau space with single hole, 10^2 if the space has two holes, 10^3 if it has three, and so on. Okay, so how many holes might there be? American superstring theorist Joe Polchinski asked a few mathematicians, and was advised that the maximum number could be 'on the order of five hundred'.[3]

A crude estimate puts the number of grains of sand on every beach at around 7.5×10^{18}. That's a lot. But now we have a potential maximum of 10^{500} different possible theories which give rise to universes with a positive cosmological constant. A few years later it was discovered that there is a potentially infinite number of theories consistent with a small negative cosmological constant.[4] And we have absolutely no means to choose between these alternatives.

Strominger had already shown in 1986 that there is a vast number of possible consistent superstring theories, many more than can be admitted if only Calabi–Yau spaces are considered, resulting in a dramatic loss of predictive power. So, this discovery was hardly new, but the realization only appears to have properly dawned on the string theory community in 2003.

I really do believe there was a time in the history of theoretical physics when these kinds of results would have been taken as evidence that a research programme had failed. I'm not alone. In 2006, Smolin wrote:[5]

> This is painful for many who have invested years and even decades of their working lives in string theory. It is painful for me, having devoted a certain amount of time to the effort, I can only imagine how some of my friends who have staked their whole careers on string theory must feel. Still, even if it hurts

like hell, acknowledging the *reduction ad absurdum* seems a rational and honest response to the situation.

With no compelling physical argument pointing to the one structure that uniquely describes our own universe, and the physical laws and the particles that we observe, you might be tempted to think that we have nowhere else to go. Time to go back to the drawing board.

But this is not what happened. Scientists can be remarkably stubborn, unwilling to let go of cherished notions, or beautifully consistent mathematical structures, clinging on when all the evidence appears to be ranged against them. Finding themselves in a deep hole, many will cheerfully reach for a bigger shovel and keep digging.

One alternative to the slow-roll inflation applied in the standard model of Big Bang cosmology was developed in 1983 by Russian theorist Alexander Vilenkin and further elaborated in 1986 by Russian-born theorist Andrei Linde. This is now commonly referred to as *eternal inflation*. In this model our universe is merely one of countless 'bubbles' of inflated spacetime, triggered by quantum fluctuations in a vast inflaton field (or fields). In certain circumstances the bubbles proliferate like a virus or like the bubbles in a bottle of champagne when the cork is popped. Such an 'inflationary multiverse' could be essentially eternal, with no beginning or end. In the multiverse, anything is possible. The essential randomness of the quantum fluctuations that trigger bubbles of inflation imply a continuum of universes with different sets of initial conditions, giving rise to different physical laws (different cosmological constants, for example).

Far from accepting the possibility of 10^{500} or an infinite number of theories as evidence for the failure of the superstring programme, some (though certainly not all) theorists instead used it to argue that all these different instances of the theory are actually describing a multiverse. To this they add some marvellously circular reasoning called the cosmological anthropic principle. Yes, there is an indescribable number of different universes out there, but we shouldn't really be too surprised to discover that we happen to find ourselves living in a universe with just the Goldilocks combination of physical laws and spectrum of particles which supports our form of life. Because if it didn't we wouldn't be here. And *voila*, yet again there's no reason to give up on superstring theory, just yet.

The ball was set firmly rolling by Stanford theorist Leonard Susskind, in a 2003 paper titled 'The Anthropic Landscape of String Theory',[6] now more commonly referred to as the *cosmic landscape*.*

Swedish-American cosmologist Max Tegmark calls this the Level II multiverse. In his hierarchy, Level III adds the many worlds interpretation of quantum mechanics (Gwyneth Paltrow's cat is alive in one set of universes and dead in another). Level IV is the multiverse described by all possible *mathematical* prescriptions.[7]

It would seems that these theorists have set sail for the metaphysical whirlpool of Charybdis with some determination, and they have then abandoned ship. The result surely cannot be, as they hope, a theory of everything, but more likely a theory of *anything at all*.

I need to emphasize that I have the utmost respect for the capabilities of the theorists involved in this enterprise, some of whom are undoubtedly possessed of the best minds of their generation. Until 2012, their tragedy was that the Nobel Prize committee observes fairly strict demands for *empirical* verification

* Smolin believes that he was the first to introduce the term 'landscape' into cosmology, based on arguments related to the 'fitness landscape' of his alternative theory of cosmological natural selection. We'll take a closer look at this in the next chapter.

of theoretical ideas before recognizing achievements through the award of a Nobel Prize. English theorist Peter Higgs and Belgian François Englert were obliged to wait 49 years before their efforts—first published in 1964—were recognized by the Nobel Prize committee following the discovery of the Higgs boson at CERN.* For as long as string theory fails to make predictions that can't simply be changed whenever the experiments or observations come up empty, the theorists will be denied Nobel-level recognition for their efforts. The prize is not awarded for theories that 'still might be true'.[8]

In July 2012, Russian entrepreneur Yuri Milner changed all that. Milner is a former particle theorist who had studied at Moscow State University and the Lebedev Institute. Having been rather disappointed in himself as a physicist, he chose to help recognize the achievements of others by using part of his fortune to establish an annual $3 million 'Breakthrough Prize', and among the first recipients of the inaugural prize for fundamental physics were many leading superstring theorists, including Witten. Linde was awarded a fundamental physics prize for—in part— developing the theory of the inflationary multiverse.

To be fair, the prize has also acknowledged the important role of experiment—prizes have since been awarded to the collaborations involved in the discovery of the Higgs boson (2013), the supernova projects that discovered the accelerating expansion of the universe (2015), the detection of neutrino oscillations (2016), and the detection of gravitational waves (2016).

Of course, Milner can spend his money however he likes. But the fundamental physics prize appears to be a rather indulgent celebration of the role of abstract theory. Prizes are awarded for

---

* Englert's Belgian colleague Robert Brout sadly died in May 2011, and the Nobel Prize is not awarded posthumously.

intellectual 'achievement', whether or not there's even the remotest likelihood of empirical verification or falsification of these theories. The awards ceremonies are lavish, featuring many celebrities from the worlds of business and entertainment, all helping to build the myth of the 'rock star theorist'.

I'd respectfully suggest that if the purpose of such philanthropy is to encourage recognition and support progress in theoretical physics, then founding an institute or funding postgraduate studentships and post-doctoral positions might be a more enlightened way to go.

In 2014, Andrew Strominger and Iranian-American Cumrun Vafa were awarded a Physics Frontier Prize in part for their 1996 efforts in using superstring theory to derive a famous result—the Bekenstein–Hawking formula for the entropy of a black hole.[9] Such is the significance of this formula to contemporary theoretical physics and the search for a quantum theory of gravity that it's worth a short detour to discover how it arises and what it implies.

In a moment of inspiration one night in November 1970, Hawking realized that the event horizon of a black hole could never shrink, meaning that its surface area could never decrease. If a black hole consumes an amount of in-falling material, then its surface area must increase by a proportional amount. And, as we think we know, what goes into a black hole never comes out.

There's another well-known physical property that in a spontaneous change can never decrease. It's called entropy, and this behaviour is the basis for the second law of thermodynamics. Could the surface area of a black hole be a measure of its entropy? Israeli theorist Jacob Bekenstein thought so and said as much in his 1972 Princeton Ph.D. thesis. But he was shouted down from all sides. For one thing, in thermodynamics the entropy of an object is inversely related to its temperature, and an object with a temperature must emit radiation.* This made no sense at all. How could a black hole possess a temperature and *emit* radiation?

Hawking set out to refute Bekenstein's hypothesis a few years later. Lacking a fully fledged quantum theory of gravity, he chose to approach the problem using general relativity to describe the black hole itself and applied quantum field theory to the curved spacetime around the event horizon. What he found was profoundly shocking:[10]

> However, when I did the calculation, I found, to my surprise and annoyance, that even nonrotating black holes should apparently create and emit particles at a steady rate. At first I thought that this emission indicated that one of the approximations I had used was not valid. I was afraid that if Bekenstein found out about it, he would use it as a further argument to support his ideas about the entropy of black holes, which I still did not like.

Bekenstein had been right all along. The entropy (symbol S) of a black hole is proportional to its surface area (A), according to an almost absurdly simple formula: S is proportional to $\frac{1}{4}A$.[11] In theory, at least, a black hole does have a temperature,[†] and emits what has since become known as *Hawking radiation*.

When matter falls into a black hole, its surface area increases, and so we must conclude that its entropy increases. But its temperature *falls*. This doesn't seem very rational—adding mass (energy) to the black hole cools it down! This is like stoking a furnace which gets colder every time we add another shovelful of coal.

---

* This is the basis of thermal imaging.
† Although this *is* very small. A black hole with a mass equivalent to that of the Sun is expected to have a temperature just 60 billionths of a degree above absolute zero. It would actually absorb more cosmic microwave background radiation than it emits.

It gets worse. If the black hole emits Hawking radiation, then kind of by definition it must *lose* energy (and hence mass) and, despite Hawking's earlier conclusion, when left to its own devices over time its surface area must therefore *decrease*. The consequent reduction in the entropy of the black hole is more than compensated by the entropy of the emitted radiation so, overall, entropy increases and the second law of thermodynamics is safe. But as the entropy of the black hole falls, its temperature rises, increasing the rate of emission of the Hawking radiation. The process snowballs and the black hole eventually 'evaporates', disappearing altogether in an explosion.

And this leaves us with a big problem.

Loosely speaking, entropy is a measure of 'disorder', related to the number of different ways in which the microscopic constituents of a complex system can be combined or distributed. In fact, entropy is related to the *logarithm* of the probability that the complex system is distributed in a certain way, according to a famous equation that is carved on the gravestone of its discoverer, Austrian physicist Ludwig Boltzmann, who committed suicide in 1906.

The second law then states that in a spontaneous change the microscopic constituents of a system 'expand' to occupy a greater number of the available states. Fortunately for me, the air molecules in the room in which I'm writing these words expand to occupy all the available space (a situation which has a high probability). They don't spontaneously accumulate in one corner, causing me to die from asphyxiation (which is not actually impossible, but has a very low probability).

In 1948 the American mathematician and engineer Claude Shannon discovered that 'information' is similarly related to the logarithm of a probability, suggesting a rather direct connection between information and entropy. Shannon was interested in the efficiency of information communicated via telegraphy, but his result is general and applies to information in all possible forms, including information 'coded' in the quantum wavefunction of any material falling into a black hole.

But what then happens when the black hole evaporates? We might be tempted to conclude that the information is then irretrievably lost from the universe. This is a problem because quantum mechanics demands that all the information coded in the wavefunction present at the start of a quantum process must be preserved at the end. We can think of this as a kind of *conservation of probability*. The system may change and the information may take different forms, but it must still be in there. So, the irretrievable loss of information implied by black hole evaporation directly contradicts the very basis and structure of quantum theory. This is called the *black hole information paradox*.

Now, of course, the Bekenstein–Hawking formula was derived using a 'semi-classical' approach: in this case general relativity and (separately) quantum field theory. We know this approach will not work satisfactorily as we push the black hole to extremes. For example, as a black hole evaporates, its surface area shrinks and will eventually reach a size at which its *quantum* properties should really be taken into account. As we know well enough by now, general relativity on its own can't deal with this.

So, we can expect that any fully fledged quantum theory of gravity should reproduce this formula, at least as a first approximation. And, with any luck, it might also resolve the information paradox.

Although the Strominger–Vafa derivation is acknowledged to be mathematically reliable and robust, it is based on a rather artificial class of so-called 'extremal' black holes in which the effects of electrical charge are as important as the gravitational effects of mass-energy. But the surface area of a three-dimensional extremal black hole can, in principle, shrink to zero, taking the entropy along with it (with all that this implies). In the Strominger–Vafa derivation the only way to prevent this from happening is to conjure up an extra spatial dimension, by 'hiding' only five of the six extra dimensions demanded by string theory, leaving one that can be used to preserve the entropy. It is also necessary to assume that the interaction strengths of the strings is precisely zero. And, of course, any application of string theory requires the assumption of supersymmetry. As Conlon states in *Why String Theory?*: 'The world in which the calculations were done looked nothing like the real world of observations.'[12]

Much more work has been done since the original derivation. The Bekenstein–Hawking formula is only the leading term in a more complicated expression for the entropy, and the fact that string theory successfully reproduces further terms is encouraging, 'even if so far only for black holes that exist in mathematical universes'.[13]

So, what does LQG predict?

Theoretical physicists (and, sometimes, popular science writers) often receive unsolicited emails full of ideas about why quantum mechanics is all wrong, how the EPR paradox can be resolved, the true nature of dark energy, or how to derive a new theory of everything. Some might set out to explain to you why your most recent paper or article is wrong. A few sentences are usually enough to set eyes rolling and direct fingers towards the 'delete' key. But, after responding to a request for copies of some of his papers, in 1994 Rovelli was surprised to receive a note from Kirill Krasnov, a 21-year-old graduate student based at the Bogolyubov Institute for Theoretical Physics in Kiev, Ukraine. Krasnov had found a small mistake in one of them.

Krasnov had become absorbed by loop quantum gravity and, with nobody at the institute able to advise him, he had been obliged to work out how to apply the theory all on his own, based on what he could learn directly from published papers. Guided by what he read, which 'allow[ed] me not to invent the invented things', he had figured out how to couple fermions to the gravitational field using techniques he had developed for himself.[14] Rovelli marvelled at his grasp of the subject and his creativity.

There followed a lively correspondence. Krasnov met Rovelli, Smolin, and other members of the LQG community at a conference in Warsaw in May 1995. He made quite an impression.

Inspired by Smolin's papers, on his return Krasnov continued his correspondence with Rovelli, as both became preoccupied by the challenge of applying LQG techniques to calculate black hole entropy. They had both realized that LQG provides a rather straightforward connection between entropy and area. The entropy is given by the logarithm of the number of different possible quantum states of a system, and in LQG the links in the spin network relate to the quantum states of space, governed by the quantum of area, which is in turn related to the square of the Planck length.

It is the links (area) rather than the nodes (volume) of the spin network that are important because the links *puncture* the black hole event horizon (the links 'disappear' beneath the horizon). Each puncture then endows the horizon with an area determined by the spin quantum number associated with the link (see Figure 28). The entropy is then simply the logarithm of the total number of different ways to endow the surface with a given area. Krasnov called it the 'geometrical entropy'.

But the result didn't come out correctly. The Bekenstein–Hawking formula states that entropy is directly proportional to area, but in this first attempt Krasnov and Rovelli arrived at a formula which featured the *square root* of the area. Something was not quite right.

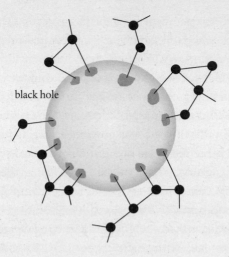

black hole

Figure 28. Krasnov developed an approach to calculating the entropy of a black hole by counting the number of punctures of its surface by links in a spin network. Each puncture endows the surface with a certain number of quanta of area, and the entropy is simply the logarithm of the total number of different ways to endow the surface with a given area.

In March 1996, whilst attending another conference in Warsaw, Krasnov had a further insight which allowed him to deduce the correct form of the relationship, and he posted a paper to the preprint archive.* Rovelli was excited. He immediately saw ways to improve on Krasnov's calculation and offered to publish a joint paper. But Krasnov's understanding of the problem was still evolving, and he refused. Rovelli was rather taken aback and concerned as Krasnov's ideas had been the inspiration for his own work. So it transpired that both Krasnov and Rovelli separately published LQG-based derivations of the Bekenstein–Hawking formula at around the same time.[15] But although both derivations

---

* At this time, the electronic preprint archive was hosted by the Los Alamos National Laboratory. It migrated to Cornell University and became known as the arXiv in 2001. Thanks to Ted Jacobson for pointing this out to me.

now had the correct form for the relationship between entropy and area, they were still somewhat troublesome and unsatisfying.

Krasnov continued to puzzle over the problem, drawing inspiration from Smolin's papers and his correspondence with both Rovelli and a senior colleague in Kiev, Yuri Shtanov. He posted a second paper on black hole entropy to the archive in May 1996. By this time he was growing increasingly concerned by what he perceived to be the imminent collapse of science in post-Soviet Ukraine. With Ashtekar's support, in the summer he moved to the Center for Gravitational Physics and Geometry at Penn State to study for a Ph.D.

In his 'linking' paper, Smolin had developed and applied techniques to calculate the entropy at a *cosmological* horizon, by 'coding' the punctures through a sample of the horizon in a boundary term. But he had overlooked the possibility of applying these same techniques to a black hole horizon. Seeking to resolve the difficulties in both his and Rovelli's derivations, by January 1997 Krasnov had understood how Smolin's approach could be applied. 'I recall my excitement when Kirill explained this to me,' Smolin later admitted, though he also kicked himself for his lack of foresight.[16]

Krasnov revised his second paper. His derivation now included the Barbero–Immirzi parameter, γ, which had by this time been introduced into the expression for the LQG area spectrum. It differed from the Bekenstein–Hawking formula: S is proportional to $\frac{1}{4}A$, by a factor $\ln 5/\pi\gamma 2\sqrt{2}$.[17] Obviously, the formula can be recovered by assuming that this factor is equal to 1, or $\gamma = \ln 5/\pi 2\sqrt{2}$, which has a value of about 0.1811.

Krasnov talked to Ashtekar and others at Penn State about his work and gave a seminar, and these ideas were added to a mix that had been simmering within the group for some years, many of which turned out to be incorrect or irrelevant. In October 1997,

Ashtekar, Baez, Corichi, and Krasnov posted a paper in which they used these techniques to show that the entropy S of a non-rotating black hole is proportional to $\frac{1}{4}A$ multiplied by the ratio γ_0/γ, where γ_0 is another constant.[18] They deduced that γ_0 is equal to $\ln2/\pi\sqrt{3}$ (which has a value of about 0.1274). Subsequent work by Krzysztof Meissner in 2004 revised this to a value of 0.23753... .[19]

Once again, the Bekenstein–Hawking formula could be recovered only by assuming $\gamma = \gamma_0$. This was rather unsatisfactory. The theorists preferred not to have to assume any such thing. The paper had taken some years to write, but it created a coherent conceptual framework, supported by detailed mathematics. But at least there was something clear to improve on or to replace. And their work had provided a handle on the size of the Barbero–Immirzi parameter, if only through another theoretical relationship. We can't test this value by reference to observation or experiment, but Ashtekar and his colleagues demonstrated that the *same value* applies for all kinds of black hole: nonrotating, rotating, uncharged, charged, and so on.

This strongly suggested that the Bekenstein–Hawking formula really ought to be completely independent of the choice of parameter. Some years later Italian theorist Eugenio Bianchi working then at the Perimeter Institute proved precisely this by focusing the derivation on fixed energy rather than fixed area, such that the Barbero–Immirzi parameter neatly cancels, leaving only S proportional to $\frac{1}{4}A$.[20] Not everyone in the LQG community agrees with Bianchi's approach, as Rovelli observes: 'It is a paper that has disturbed many, because everybody had his own little way of viewing the issue, and Eugenio cuts through everything, so it has raised much defensive resistance, but it is a masterpiece.'[21] Smolin agrees.

One more thing before we move on. As Smolin had observed in his 'linking' paper, so Ashtekar and his colleagues noted that the quantum states which dominate the procedure they had used to count all the possible punctures in the event horizon correspond to links in the spin network with values of the spin quantum number of ½. We know that any object with a spin quantum number of ½ may have two possible 'orientations', corresponding to +½ and −½. Such 'up'/'down', or 'on'/'off', or 'yes'/'no' possibilities are said to be *binary*. They correlate with the binary 'bits' of information used in computing, which are taken to be 0 and 1.

As is often found to be the case in this business, John Wheeler was there first:[22]

> Trying to wrap my brain around this idea of information theory as the basis of existence, I came up with the phrase 'it from bit.' The universe and all that it contains ('it') may arise from the myriad yes-no choices of measurement (the 'bits')…Information may not be just what we *learn* about the world. It may be what *makes* the world.

Wheeler speculated that the area of a black hole event horizon is 'tiled' by quanta of space on the order of the square of the Planck length, each tile representing a 'bit' of information that is 'shielded' by the horizon (Figure 29).* Ashtekar and his colleagues concluded: 'Thus, there is a curious similarity between our detailed results and John Wheeler's "It from Bit" picture of the origin of black hole entropy.'[23]

Of course, elevating the status of 'information' to a fundamental description of material reality has a few consequences. I think you can at least understand why theorists are exercised by the black hole information paradox.

Intellectual battles between rival theorists make for good reading, and Hawking always enjoyed the odd wager. Hawking

---

* Strictly speaking, as these are quantum bits (or 'qubits', for short), then Wheeler's aphorism is really 'it from qubit'.

Figure 29. Wheeler had earlier imagined that the surface of a black hole is 'tiled' by quanta of space which represent 'bits' (or 'qubits') of information.

insisted that quantum information is indeed irretrievably lost inside an evaporating black hole, and in 1997 agreed to a bet with theorists John Preskill and Kip Thorne, who argued the opposite. Susskind and Dutch theorist Gerard 't Hooft also believed that information should be somehow preserved. The resulting battle of wits is recounted by Susskind in his 2008 book *The Black Hole War*.

If the scrambled bits of information are not to be lost forever, then either they are somehow preserved on the surface, to be eventually emitted in the form of Hawking radiation, or they are preserved in some kind of remnant left behind after the black hole has evaporated completely. To Susskind, the latter seemed unlikely, so he pitched for the former.

Following an almost throwaway remark by 't Hooft, made during a visit in 1994, Susskind realized that 'the maximum amount of information that can possibly be contained in any region of space cannot be greater than what can be stored on the boundary of the region'.[24] He called it the *holographic principle*.

We're already come across similar relationships in the general boundary formulation of quantum field theory described in Chapter 11.

The holographic principle was no more than an idea, one without a grounding in formal theory. But in 1998, Argentinian theorist Juan Maldacena announced a powerful new conjecture. He deduced that the physics described by a certain type of super-string theory in an n-dimensional spacetime is equivalent to the physics described by a supersymmetric quantum field theory applied to its $(n-1)$-dimensional boundary.

This result stirred up a lot of excitement in the string theory community. A superstring theory which implicitly includes quantum gravity is equivalent (it is said to be 'dual') to a super-symmetric quantum field theory in a fixed background space-time, which doesn't include gravity. Witten went on to show that a black hole in the bulk spacetime of the superstring theory is equivalent to a hot 'soup' of elementary particles, such as gluons, on the boundary surface.

On seeing Witten's paper, Susskind understood that the black hole war had finally been won. He wrote:[25]

> Quantum Field Theory is a special case of Quantum Mechanics, and information in Quantum Mechanics can never be destroyed. Whatever else Maldacena and Witten had done, they had proved beyond any shadow of a doubt that information would never be lost behind a black hole horizon. The string theorists would understand this immediately; the relativists would take longer. But the war was over.

Any information passed to the interior of a black hole can still be recovered from the dual quantum field theory description, to be dissipated (but not lost) in the Hawking radiation.

But there are, perhaps inevitably, more than a few caveats. The duality encapsulated in Maldacena's conjecture involves a superstring theory in something called anti-de Sitter space, named for Dutch physicist Willem de Sitter who, in 1917, solved Einstein's gravitational field equations for a model universe empty of matter in which spacetime expands exponentially. We can think of such a universe as consisting only of dark energy, with a positive cosmological constant, and which therefore has positive spacetime curvature. As on the surface of a sphere, the angles of a triangle drawn in de Sitter space will add up to more than 180°.

In an anti-de Sitter space, the cosmological constant is *negative*, the spacetime curvature is negative, and the angles of a triangle add up to less than 180°. This is a hyperbolic universe shaped like a saddle. Inject matter into an anti-de Sitter universe and the curvature of spacetime causes it to be pushed away from the boundary and drawn towards the centre. For this reason Maldacena's conjecture is sometimes referred to as the AdS/CFT duality, where AdS stands for anti-de Sitter and CFT stands for conformal field theory, which is a special class of quantum field theory.

We should also acknowledge that, even today, Maldacena's conjecture is just this—a conjecture. This duality between a specific kind of superstring theory (for which we have no evidence) and a supersymmetric quantum field theory (for which we have no evidence) is itself not proven. In his book *Why String Theory?*, Conlon accepts that: 'Neither side of the correspondence involves an object with a rigorous mathematical definition.'[26] The correspondence has been checked many, many times 'by calculation' but this does not constitute a formal proof. Some string theorists have declared that it belongs to a category of things that are 'true but not proven'.[27]

Hawking didn't concede immediately. He figured out his own reasons why he was wrong or, as he reported to a conference in

Dublin in 2004, 'everyone was right in a way'.[28] Thorne was still doubtful, but Preskill accepted payment, in the form of a baseball encyclopaedia, 'from which information can be recovered with ease'. However, despite this and Susskind's claims, the black hole war doesn't seem to be over. String theorists such as Polchinski have sought to refute Susskind's arguments, questioning the various postulates which form their basis.[29]

Perhaps, after all, there's a better way to try to understand what's going on here. Could it be that the black hole information paradox is really an artefact of the application of general relativity—a theory in which the spacetime that emerges is assumed to be *continuous*—in a universe that is fundamentally quantum in nature? Could this be nothing more than Zeno's paradox, based on rather more exotic objects, dressed up in rather more exotic language?

We saw in the previous chapter how LQC eliminates the Big Bang singularity for certain types of model universes. In LQG, singularities are excluded, for the simple reason that there can be no area smaller than a single quantum of area; no volume smaller than a single quantum of volume, all of which is a bit like saying there can be no quantity of light smaller than a single photon. The effects of quantum gravity in LQC, particularly the quantum repulsive force that causes the bounce, become important when the *curvature of spacetime* reaches the Planck scale. This can happen long before the radius of the universe itself shrinks to the Planck length. Applying the same logic implies that quantum effects will arise long before the radius of a collapsing star reaches the Planck length.

This doesn't necessarily mean that a black hole cannot evaporate completely. It may undergo a bounce, as we will consider below. Or, instead of a singularity, it may leave a quantum region which evaporates very slowly (which is what Ashtekar suspects).[30]

In 2014, Rovelli and Vidotto, now working at the University of Nijmegen in the Netherlands, suggested that a black hole doesn't collapse to a singularity at its core, but rather collapses to what they called a *Planck star*.[31] A black hole with a mass equal to that of the Sun would then have a Planck star at its core with a radius of about 10^{-10} centimetres. For sure, this is highly compressed—the mass is squeezed into a volume smaller than an atom—but this is still thirty orders of magnitude larger than the Planck length. As the black hole evaporates, it is likely to develop a *very* long neck, making it possible to house all the quantum information inside the horizon, so preventing it from being lost.

But rather than collapse completely to nothing, the black hole core bounces as the effects of quantum gravity kick in. To an observer somehow able to survive sitting on the surface of the Planck star, this bounce occurs very quickly. But to an external observer the effects of gravitational time dilation mean that the bounce takes longer than the present age of the universe.

There are now two horizons. The outer event horizon shrinks as it emits Hawking radiation and the black hole evaporates. But the same mechanism causes the inner horizon of the Planck star to expand. Eventually, the two horizons meet, at which point any residual information that has not radiated from the event horizon can escape. Information is not irretrievably lost from the universe and the paradox is resolved.

Black holes can, of course, be formed from stars much smaller than the Sun. Any black holes formed in the very early universe with masses around 10^{12} kilograms* should be evaporating and releasing the energy from their Planck stars about now. The quantum-gravitational signature of the death of such a primordial black hole would be a short burst of intense, high-energy

---

* The mass of the Sun is around 2×10^{30} kilograms.

radiation, as its Planck star releases its energy and flashes out of existence.

Taken together, signals characteristic of very short gamma ray bursts (VSGRBs) captured from seven different satellite-borne detectors hint at the possibility that they may originate from such primordial black holes.[32] One possible 'smoking gun' would be observation of a characteristically flattened redshift-distance curve: primordial black holes exploding now would be necessarily smaller, emitting higher frequency radiation which partly compensates for the cosmological redshift.

For sure, we're a long way from determining whether these bursts of radiation can be ascribed to Planck stars, if indeed these exist, but at least this proposed solution to the information paradox might eventually give us something to look for, instead of something merely to speculate about.

16

CLOSE TO THE EDGE
The Reality of Time and the Principles of the Open Future

'Daddy,' Lee Smolin's son asked, 'did you have my name when you were my age?'[1]

Young children go through a phase in which they express boundless curiosity about the world around them, asking endless questions of their parents in an attempt to understand what they're experiencing. Over time, this expands into a broader curiosity about where they fit in the general scheme of things. Sometimes their questions leave us wishing we had paid more attention in biology class. Other questions might lead us to moments of philosophical reflection, as deep as they are personal.

His son's perfectly innocent question pulled Smolin up short. The mathematical constructions that are bread and butter to the theory of quantum gravity lead us inexorably to the conclusion that time is somewhat illusory. We are obliged to presume that it is an emergent, relational property, assembled from a causally connected sequence of frozen instants, like the ribbon of still pictures projected in front of our eyes so rapidly they create the illusion we call a movie.

And yet, freed from the burden of knowledge of any of the physics of the past four centuries, the naïve question betrays a potentially more profound truth. Smolin's son had come to understand perfectly well that there was a time before he was born, and he was curious to learn what might have happened then.

No matter what the mathematical constructions say, our *experience* of the passage of time is undeniable. As we grow older, the endless summers of youth fade in memory as whole years vanish in a riot of happenings. 'And then one day you find, ten years have got behind you'.[2] We smile at a photograph, then pause to ask ourselves: Was that *really* ten years ago (or twenty, or thirty—you choose)?

But so much of the physics of the past century proved to be completely counterintuitive, and we learned that our human experience of the large scale is no real guide to what lies beneath. As a student and young academic researcher, Smolin had come to embrace what physics was saying about the unreality of time because it suited his adolescent desires. The time-bound world of human affairs was ugly and inhospitable, and the timeless truths of physics offered the possibility of escape.

But then life happened, and the world didn't seem so bad, after all.

His experiences wrestling unsuccessfully with the task of bringing loops and strings together and completing the LQG programme led him to embrace approaches which promised much more respect for the status of time and causality, such as Markopoulou's causal spinfoams, and the theory of causal dynamical triangulations championed by German theorist Renate Loll and her colleagues. He was growing increasingly concerned about both the origin and status of 'natural laws', and continued to ponder the seemingly inherent non-locality of quantum mechanics.

These influences now converged on the older, wiser Smolin. After some soul-searching, he was struck by a radical, if not revolutionary, thought. *What if time is real?*

From his brother David, a law professor at Cumberland Law School in Birmingham, Alabama, Smolin had heard about Roberto Mangabeira Unger, a Brazilian-American philosopher, legal theorist, and politician. Unger has been a professor at Harvard Law School for several decades, best known in legal circles for his contributions to the Critical Legal Studies movement in the 1970s and 1980s. In 1985 he supported Brazil's transition to democracy following more than twenty years of military dictatorship, and for two years served as Minister of Strategic Affairs during President Luiz Inácio Lula da Silva's second term in office in 2007. He was re-appointed to the same position in February 2015 by President Dilma Rousseff, just nine months before the commencement of impeachment proceedings against her.

Smolin had heard Unger lecture on legal theory when he was on the faculty at Yale. 'He was dazzling.'[3] It emerged that Unger held similar views about time, and some years later they agreed to collaborate.

The collaboration resulted in two books, a popular summary written by Smolin and published as *Time Reborn* in 2013, and a detailed, jointly authored volume titled *The Singular Universe and the Reality of Time*, aimed at a more specialized audience, published in 2015. *Time Reborn* was the inspiration for Canadian playwright Hannah Moscovitch's play *Infinity*, first performed on stage in Toronto in March 2015, and on which Smolin collaborated.

Writing is one of Smolin's preferred methods for working through the hard puzzles that confront him as a professional scientist. But, by his own admission, *Time Reborn* was 'quite frightening to write'.[4]

To appreciate what makes this idea so radical it's helpful first to understand just how physics managed to lose its grip on time so completely. In the opening chapters of *Time Reborn*, Smolin traces this history back to Galileo. It's perhaps no real surprise that the beginning of the fall of time coincides with the rise of measurement and particularly mathematics as the principal language we use to describe nature.

In 1638, whilst under house-arrest, Galileo published *Discourses and Mathematical Demonstrations Concerning Two New Sciences Pertaining to Mechanics and Local Motions*. In the section on the motions of projectiles, he considers an object (such as a cannonball, fired from a castle parapet) moving with uniform speed horizontally through space. This motion is then compounded with the accelerated downward motion due to gravity, and as the object moves forward it also falls to the ground. The question Galileo wanted to address is this: What is the *trajectory* of the object through space?

In Theorem I, Proposition I, he writes:[5]

> In like manner it may be shown that, if we take equal time-intervals of any size whatever, and if we imagine the [object] to be carried by a similar compound motion, the positions of this [object], at the ends of these time-intervals, will lie on one and the same parabola.

We know all this, of course. The uniform horizontal motion covers equal distances in equal times t but the object falls due to the action of gravity through distances which increase with t^2, the square of these same time intervals. The result is a parabola.

This all seems perfectly reasonable and we might wonder why we're making a fuss. But think about the game that's being played here. Galileo took elements of physical reality—space and time—and turned them into elements of *geometry*, as we can see from Figure 30. This might be fine when it comes to space, because at least there is some sense of permanence about space—whatever is *in* space might come or go, but in classical Newtonian mechanics space itself doesn't change in any fundamental way. But this is not at all true for time. Time is impermanent. It changes. When we record a sequence of measurements made over a number of time intervals, we necessarily *freeze* them for later inspection. As Smolin explains: 'The method of freezing time has worked so well that most physicists are unaware that a trick has been played on their understanding of nature.'[6]

Well, okay, we accept that the game involves an abstraction, and our familiarity with it may mean that we've lost sight of its

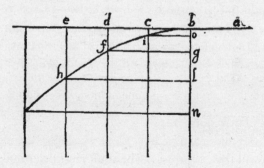

Figure 30. As this illustration from Galileo's *Two New Sciences* shows, he took elements of physical reality—space and time—and turned them into elements of geometry, mapping uniform intervals of time to a horizontal axis and spatial distances (heights) to a vertical axis. He wrote: 'Draw the line *be* along the plane *ba* to represent the flow, or measure, of time; divide this line into a number of segments, *bc*, *cd*, *de*, representing equal intervals of time'.

nature and extent. But, just as surely, time is still very much part of the picture. If we make use of the coordinate system devised by Descartes, we map three-dimensional space onto x, y, and z coordinates. But we also have a t coordinate: time might have become an aspect of geometry, enslaved to measurement and mathematics, but it hasn't exactly disappeared.

Yes, we still have time as part of our *representation* but, make no mistake, our endgame here is the set of underlying laws that determine the shape of the trajectory in time and space. We will come to know these as Newton's laws of motion and universal gravitation. These laws are *timeless*. Their representation might contain t but they are *laws*. They are supposed to stand, immutable, for all eternity.

Fast-forward a few hundred years. Newton's absolute space and time are displaced by Einstein's relativity, and the notion of simultaneity loses its meaning. Minkowski dooms space and time to fade away into mere shadows, to be replaced by a space-time in which time has become just another dimension and the universe is as unchanging as a block of stone.[7] Einstein turns spacetime into curved geometry. Then, as we learned in Chapter 6, a conclusion from the earliest attempts to develop a canonical quantum theory of gravity is that time is replaced by a connected sequence of frozen moments.

To recover physical time in LQG we must choose how to conjure it from such moments. We assume that time emerges from the dynamical evolution of spin networks, which we call spinfoam. The illusion of time emerges from changing relationships between the atoms of space. In LQC we assign responsibility to an arbitrary field as a kind of proxy for time, or instead we pursue the development of a spinfoam cosmology.

All this, argues Smolin, is the inevitable consequence of a belief that the purpose of physics is to uncover the eternal truths of the universe, translated into a search for the single, timeless mathematical law, equation, or theory that will tell us where the universe comes from and how it works. 'This is such a familiar habit of thought that we fail to see its absurdity,' says Smolin. If it is to describe every aspect of the universe, such an equation would necessarily have to be established 'outside' it. We're right back with the 'God's eye view' that Einstein sought to banish by bringing our rulers and our clocks *inside* the universe. 'If the universe is all that exists, then how can something exist outside it for it to be described by?'[8]

It's easy to forget that the mechanical philosophers who worked to establish the foundations of our scientific worldview in the seventeenth and eighteenth centuries believed quite passionately that their purpose was to discover the intricate details of God's design. Today we might politely overlook the extended discussion of the nature of God that Newton included in the general scholium added to the end of Book III in the 1713 second edition of the *Mathematical Principles*: 'And this much concerning God; to discourse of whom from the appearances of things, does certainly belong to Natural Philosophy.'[9]

Well, okay, but that was three hundred years ago and surely science has moved on? But then we remember that Einstein's remark about playing dice wasn't idle, but was founded in his belief in God, albeit a God much less biblical. Hawking concludes *A Brief History of Time*, with the remark: 'If we find the answer to [a complete theory], it would be the ultimate triumph of human reason—for then we would know the mind of God.'[10] In *Dreams of a Final Theory*, Stephen Weinberg devotes a whole chapter to God, observing: 'If there were anything we could discover in nature that *would* give us some special insight into the handiwork of God, it would have to be the final laws of nature.'[11] I should point out that Weinberg goes on to conclude: 'Judging from this historical experience, I would guess that, though we shall find beauty in the final laws of nature, we shall find…no hint of any God who cares about such things. We may find these things elsewhere, but not in the laws of nature.'[12]

Few theoretical physicists today might openly admit that they're searching for God (at least, not until they have established their scientific reputations or won a Nobel Prize). But most will happily admit that they're engaged in what I can only call a *spiritual* quest for eternal truths that are greater than themselves, and greater even than the universe. It's stuff like this that makes us human, of course, but we should be clear that embarking on such a quest requires what Unger calls a 'metaphysical pre-commitment'.[13]

Knowingly or not, such a pre-commitment has led some theorists to ponder on the 'unreasonable effectiveness of mathematics in the natural sciences',[14] which is all too easily translated into a sentimental longing for mathematical *beauty* in physics. At its extreme, this becomes Tegmark's Mathematical Universe Hypothesis—reality is timeless mathematics. This kind of thing exalts mathematics: it puts it up on a pedestal, and we lose sight of what I (for one) believe it really is: a very powerful and effective language *invented by humans* to describe and discover patterns in nature. When we perceive beauty in the mathematics, I think what we're really perceiving is an underlying beauty in nature itself.

This goes deep. Smolin and Unger are challenging the rules of the game, the very basis on which we've made nearly four hundred years of progress in physics. They're calling for a very different kind of metaphysical pre-commitment, one that they hope will spark a new revolution in the science of physical cosmology.

Now, revolutions are fought on the basis of ideas or ideals or what, ever since the publication of American philosopher

Thomas Kuhn's *The Structure of Scientific Revolutions*, we've tended to call 'paradigms'. In a manner of speaking, the whole LQG programme is revolutionary (perhaps with a small 'r'). Or at least it was in the 1980s and 1990s when the foundations of the theory were being laid. Typically, when the fighting is over and some kind of victory is declared, it becomes necessary to set about the task of establishing the new order that is to be built on these foundations. This means coming to terms with the implications of the new paradigm, solving the plethora of smaller problems in order to set enduring structures in place. This requires what Kuhn referred to as 'normal' science, a commitment to puzzle-solving that demands pragmatism and doggedness.

Having stepped down from the barricades, Rovelli embraced the challenge of establishing some sense of order in the messy kitchen of quantum gravity. 'I like washing dishes,' he told British journalist Bryan Appleyard. 'One starts with a big mess, focuses on it without thinking of anything else, the mind goes free and, after a while, everything is neat and clean.'[15]

This is why it was necessary for Rovelli to clarify his position on the interpretation of quantum mechanics, described in Chapter 13. If quantum mechanics is somehow incomplete, how can it possibly serve as a stable foundation on which to construct a quantum theory of gravity? To have any realistic chance of establishing a theory that could endure, Rovelli first had to avoid getting tangled in endless debates about the nature of quantum reality and the completeness or otherwise of quantum mechanics. He needed to wash the dishes.

One result of such diligence was a technical book, titled *Quantum Gravity*, published by Cambridge University Press in 2004, when Rovelli believed he had understood these foundations properly. All things considered, this book is one of his major contributions to the field: an in-depth discussion on how to think about quantum space and quantum time. The entire first half revisits elementary classical and quantum physics and general relativity, but reconsidered from a new perspective that allows a quantum theory of gravity to be formulated. It explains what it means to work with a theory in which evolution in time is not a primary concept.*

But Smolin has never really stopped fighting. He's still at the barricades. Revolutionaries are driven by a burning desire to break things down in order that better things can be built to replace them. As he prepares to throw another Molotov cocktail, Smolin urges us to rethink our understanding of the nature of time.

There are always consequences. Being a visionary means sailing very, very close to the edge of the whirlpool of metaphysical nonsense that is Charybdis. The best you can do is keep a telescope firmly trained on the rock shoal of empiricism, and hope for the best.

As he researched potential faculty members for the new Perimeter Institute, Howard Burton identified two broad categories of theoretical physicist:[16]

meticulous calculators who will not announce the slightest result unless it is checked six ways from Sunday, and wild speculators who indulge themselves by sketching out crazy theories. When someone from the first group announces a result, you can be sure it's right, but not at all certain it will be interesting. When someone from the second group announces a result, you can be certain it will be interesting, but not at all sure it will be right.

---

* Whilst this book was in production, Rovelli published a popular account of his arguments against the reality of time, called *The Order of Time*.

He was sure that Smolin is a theorist of the second kind. This doesn't mean that Smolin doesn't do the dishes (his wife Dina assures me that he does), or that he enjoys the task any less than Rovelli. But Smolin is also imbued with a restless energy, his head buzzing with novel ideas and his eye always on the questions that penetrate to the very deepest caverns of the meaning of physics.

Let's be clear that Smolin and Unger are not arguing that the laws of physics as we know them today should be rejected. Far from it. What they're saying is that we need to remember that they have a limited domain of applicability. Our mistake is to think that they can be applied to the universe as a whole. This is what they call the *cosmological fallacy*.

In fact, as we currently understand them, the laws themselves are really rather less grand than some theorists would have you believe. When we look closely, we discover that no matter how scientific they might seem, they are *ideals*. And as ideals, they can't actually ever be properly tested in isolation. Whenever we apply such a law to a real situation, we find it first necessary to simplify the conditions so that we can focus our attention on the important phenomena. This is what Smolin calls doing *physics in a box*.

It's then necessary to make some simplifying approximations or invoke what philosophers of science call *auxiliary hypotheses*. This might be as straightforward (and as obvious) as neglecting aspects of the complexity of nature in the interests of testing a simpler, more accessible model—for example, neglecting the effects of gravity when studying electromagnetism. As Smolin writes: 'Newton's first law of motion asserts that all free particles will move along straight lines. It has been tested and confirmed in numerous cases. But each test involves an approximation, because no particle is truly free.'[17]

What then gets tested is not the law itself, but the law wrapped in a blanket of auxiliary hypotheses. If an observation or experimental test comes up negative, a scientist's initial reaction is to point the finger of blame at one or more of these hypotheses, rather than question the veracity of the law (which, after all, is supposed to be immutable).[18]

This doesn't make progress impossible.[19] Doing physics in a box means isolating part of the universe in which all the interesting stuff is happening inside (according to a representation based on time intervals, t) and assuming that the rest of the universe outside behaves like a kind of timeless backdrop, a stage upon which all the actors of the physics strut and fret their hour. Smolin and Unger argue that this same strategy simply cannot be used when we try to develop a *cosmology*.

By taking time to be an illusion we've become mired in a set of cosmological conundrums. We cannot explain why the laws of physics take the forms that they have. We cannot be specific about the initial conditions that prevailed in the earliest moments of our universe, leaving us to ponder the fine-tuning of its laws with growing exasperation, or to abandon ship and reach for the string theory landscape, eternal inflation, and the multiverse.

But then how should we proceed?

Smolin and Unger adopted much the same strategy as Einstein in 1905.[20] Why not simply assume that there is only one universe, and that time is real, and explore the consequences? How difficult could it be?

At least one thing is crystal clear. 'If, on the other hand, time is truly real,' Smolin writes, quoting Unger, 'then nothing, not even the laws, can last forever… They must *evolve*.'[21]

Time comes before law.

Perhaps surprisingly, it's not all that difficult to imagine circumstances in which the laws of physics might change. Eliminating the Big Bang singularity in LQC means that there may have been a universe that existed before ours. If the laws are not completely reset at the bounce, but are instead randomly tweaked, then we can perhaps understand how they might evolve over time in a universe of repeating cycles of bounce, expansion, contraction, and bounce.

But, as we've seen, such a cyclic LQC still requires the assumption of cosmic inflation, and although it takes the pressure off the problem of initial conditions it doesn't completely resolve it. There are, however, other cyclic cosmologies that can do without both Big Bang singularities *and* inflation. One such is the model first devised in 2001 by Princeton theorists Paul Steinhardt and Justin Khoury, working with Burt Ovrut at the University of Pennsylvania and Neil Turok, then at Cambridge University in England (and now at the Perimeter Institute). This is based on collisions between M-theory branes in which a small proportion of the energy of the collision is converted into hot radiation. We recognize this as a Big Bang, accelerating the expansion of space-time without the need for inflation. The branes collide and bounce apart, but never separate far enough to escape their mutual gravitational attraction. Some trillions of years later the branes collide again, the universe resets, and the cycle repeats.

As there's no need to invoke inflation in this model there will consequently be no primordial gravitational waves. But, as they say, absence of evidence is not evidence of absence—the fact that primordial gravitational waves haven't yet been found doesn't mean they won't be found in the future. The cyclic models also predict that the temperature variations in the cosmic background radiation are not entirely random and, as I mentioned in Chapter 14, reinterpretation of the most recent Planck satellite data led Steinhardt and his colleagues Anna Ijjas and Abraham Loeb to declare that the data strongly disconfirm the simplest inflationary scenarios.[22]

It's probably not unreasonable to say that the jury is still out. It's true that such cyclic cosmologies don't make much sense if time is not assumed to be real. Steinhardt and Turok argue that the cyclic model also drives the universe towards laws and initial conditions that are familiar: 'The underlying mechanism driving the cycles is gentle and self-regulating...Simplicity and parsimony reign. Virtually every patch produces galaxies, stars, planets, and life, over and over again'.[23] However, if each collision resets the laws and initial conditions in a random fashion, as we might expect, then it's difficult to see how successive cycles can give us what we need.*

What we need is a mechanical model which drives *irresistibly* towards the apparent complexity of the laws and initial conditions that prevail in our own universe. Smolin described just such a model in his first book *Life of the Cosmos*, published in 1997, based on ideas first published five years earlier. He calls it *cosmological natural selection*. It involves recycling, but not in the sense of a cyclic cosmology. Instead, the black holes that populate the universe are themselves seeds for the production of whole new regions of spacetime, or new universes.

In LQC, the progenitor of a universe is the collapsing universe that existed before the big bounce. In cosmological natural selection, the progenitor of a universe is the black hole that existed before it bounced.

A 'typical' universe is then more likely to be born to a parent that has many offspring (many black holes) rather than few. Just as the evolution of life forms on Earth is directed through the processes of random genetic mutation and natural selection, so over time the universe tends towards a structure which self-perpetuates, maximizing the number of black holes.

If each black hole bounce randomly 'mutates' the laws of physics slightly (perhaps yielding a set of elementary particles with

* In *The Singular Universe and the Reality of Time*, Smolin argues that the trajectory through each collision would need to be governed by a dynamical attractor. See p. 453.

slightly different masses and forces with slightly different strengths), then cosmological natural selection leans towards universes which become self-perpetuating, tuned to maximize the production of black holes. It so happens that, based on our current understanding, this structure also requires rather familiar elementary particles, atoms, stars, galaxies, and laws which govern their properties and behaviour. 'There's a bonus: While the explanation involves maximizing the production of black holes, a consequence is to make the universe hospitable to life.'[24]

Why not cut out the middle man and just evolve to a universe which maximizes the number of *primordial* black holes? Because such universes are necessarily smaller. Universes which can produce black holes through the gravitational collapse of stars can be much larger and can therefore produce black holes in greater number.

Yes, this is fanciful speculation, the result of sailing so very close to the edge of Charybdis. What we have here is another kind of multiverse theory but, unlike the multiverse of eternal inflation and the string theory landscape, in this theory all the different universes share a similar lineage. They evolve towards a life-friendly structure so there's no need to invoke the anthropic principle.

And Smolin argues that, unlike proponents of an anthropic multiverse, he has not abandoned ship. He believes that cosmological natural selection is testable. If the universe is indeed driven to a structure which maximizes the number of black holes, then there are implications for the way that cosmic inflation must work (if it really happens), there are potential implications for the value of the cosmological constant, and we can expect a hard upper limit on the range of masses of neutron stars.

The typical mass of a neutron star is about 1.4 times the mass of the Sun, with a radius of about 11 kilometres. It consists of neutrons, protons, and electrons with at least 200 neutrons for every proton. It may also possess a core of kaons, more exotic relatives of protons and neutrons in which a down quark is replaced with a strange quark. Not all neutron stars are *pulsars*, but the vast majority are.

In a universe in which the laws of physics have been tuned by evolution to maximize the production of black holes, there is a necessary upper limit on the mass of neutron stars, which Smolin estimates to be 2 solar masses. Here, then, is a prediction. Observation of neutron stars significantly heavier than two solar masses will confound the theory of cosmological natural selection.

A recent (2013) review by James Lattimer at Stony Brook University in New York puts the largest, well-measured mass at 1.97 ± 0.04 solar masses. But there are a couple of so-called X-ray binary pulsars, PSR B1957+20 (the menacingly named 'black widow pulsar', which is consuming matter pulled from its companion star) and 4U 1700-377, both of which are reported to have masses of 2.4 solar masses. Unfortunately, there are large systematic errors inherent in mass measurements from X-ray binaries, so these data cannot be considered conclusive.[25]

If more accurate data eventually bring these masses below the 2-solar-mass threshold, nothing is really proven as such results are consistent with a number of accepted theories of nuclear physics. But if the 2.4-solar-mass results stand firm, Smolin's theory can be ruled out, and theories invoking exotic states of matter based on quarks come under pressure, too.[26] I guess we'll just have to wait and see.

Perhaps the most significant challenge to the reality of time is what Smolin and Unger refer to as the 'meta-law dilemma'. If the laws of nature evolve, doesn't this suggest a meta-law which governs their evolution? Otherwise, before the laws of nature are 'locked down', how are the elementary constituents and the forces that operate on them supposed to 'know' what to do? If there is a meta-law of nature, then we're left to ponder: Why *this* meta-law?

In Chapter 14 I explained that the lowest-energy orbital of the electron in a hydrogen atom is represented by a spherical wavefunction. This wavefunction gives a quantum statistical probability distribution for the electron which peaks at the Bohr radius. 'If you prepared a hydrogen atom fifty years ago and you looked for where the electron is you get the same statistical distribution as you do now, and that you will with confidence get in a hundred years.'[27] We conclude that Coulomb's law, which governs the electrostatic force between the proton and electron, is a law of nature which operates outside of time.

Smolin believes this is the result of overinterpreting the evidence, and instead argues for a *principle of precedence*. The operation of legal systems relies not only on regulations and laws written by the government into the statute book, but also on a body of *common law* established by judges and the courts. In a legal dispute which is unprecedented and fundamentally different from any previous case, the judge's decision establishes a precedent which guides future decisions.

It seems pretty obvious that the very first physical interactions occurring in the universe were unprecedented. If we further assume that they were governed by no predetermined laws, we can imagine a scenario in which the initial interactions are rather hit-and-miss, producing a *range* of outcomes. Relational precedents are established for both the states of the elementary constituents (whatever they may be) *and* their interactions with each other. Nature learns by doing.

A judge's common law decision is written into court proceedings, and many such decisions are collected and published in commentaries, restatements, and encyclopedias (such as the

Corpus Juris Secundum in the USA). In this way precedents established in the past influence future decisions.

By analogy, Smolin's principle of precedence therefore requires a mechanism through which the outcomes of past interactions will be 'selected' in future interactions. 'This would seem to require a new kind of interaction,' writes Smolin, 'whereby a physical system can interact with copies of itself in the past.'[28]

Figuring out how this might work resulted in a new, non-local hidden variable interpretation of quantum mechanics that Smolin calls the *real ensemble interpretation*.[29] This is so named because the states of quantum entities are assumed to be real (and not just expressions of coded information) and the entities in these states interact in concert, just like an ensemble of musicians.

Suppose we devise an experiment to determine the location of an electron in a hydrogen atom. We perform a measurement and discover the electron at a distance just a little beyond the Bohr radius (look back at Figure 26 in Chapter 14). We repeat the measurement and find that the electron is now a little way inside the Bohr radius. We might conclude that between measurements the electron moved closer to the nucleus, but in the real ensemble interpretation Smolin argues that the outcomes we get are produced by *copying* from the members of the subgroups (the ensembles) of all hydrogen atoms in the universe in which the electron is just a little beyond, and just a little way inside, the Bohr radius.

We continue making measurements, yielding outcomes that we interpret in terms of a quantum probability distribution which peaks at the Bohr radius. But what we really have is a distribution of the *numbers* of hydrogen atoms belonging to each ensemble. Throughout the universe, there are more hydrogen atoms with the electron precisely at the Bohr radius than at any other distance, so these get copied more often.

This leaves us to ponder how the highly non-local copy process might work. Of course, we already have the answer. As I explained in Chapter 13, there may be non-local links between points in the dynamic lattice that represents the emergent space of spin networks. A hydrogen atom that is subjected to a measurement is connected to other members of the ensemble to which it belongs through non-local links. Copying takes place once more 'beneath' space.

Assuming such interactions happen, the precedents are repeated again and again in what we will come to recognize as quantum statistical distributions. Building on ideas published by Lucien Hardy and Llúis Masanes, and Markus Müller, Smolin has speculated that the operation of a meta-law might then drive these repeated outcomes towards a set of simple 'rules' which serve to minimize the amount of information required to reproduce them. Over time, these rules stabilize to the laws that govern the properties and behaviour of matter, radiation, and spacetime.

If all of this is right, then by creating novel entangled quantum states in the laboratory we create a system completely without precedent. Nature has no historical precedents to draw on in order to decide what to do, and the outcomes of any measurements on such states should be utterly unpredictable. However, such an experiment would be extremely difficult to interpret, as random, unpredicted outcomes are also consistent with an apparatus that has been set up improperly. A seasoned experimentalist will prefer to repeat the measurement, ironing out any wrinkles in the setup (and usually looking for the outcomes to become more predictable and reproducible). By this time, the principle of precedence will have done its work.

There's much more. The reality of time also demands a reinterpretation of general relativity, in which the relativity of time is replaced by the relativity of physical shape. If time is real, then it may be space that is the illusion, as described by theories such as causal dynamical triangulation and quantum graphity.

This doesn't mean that Smolin has abandoned the theory he has spent so much of his career helping to develop. It does mean that he sees it only as a kind of stepping-stone, one that does not (because it cannot) resolve the deep problems associated with what he perceives to be the fundamental incompleteness of quantum mechanics; and one that does not (because it seems it cannot) yield general relativity as a limit except in terms of 'physics in a box'.

For sure, all of these real-time models are very speculative and many more questions are provoked than answers provided: 'There's much work yet to be done... before we can conclude that they may be realistic', he writes.[30] But these are early days. You don't overturn the metaphysical pre-commitments inherent in four hundred years of physics that easily.

When rational argument based on the available evidence is sufficient to make a judgement, a decision can be made. But when this is insufficient, as it is today in physical cosmology, Smolin argues that the community must encourage a wide and diverse range of hypotheses and seek evidence for these so that a decision can be possible, at least in principle.

These are what Smolin refers to as the *principles of the open future*. 'As we seek to develop this new science we will discover that our success can be measured by the extent to which the future of cosmology becomes the cosmology of the future.'[31]

EPILOGUE

Like Being Roped Together on a Mountain

In the Preface, I claimed that when it comes to our ability to comprehend the nature of space and time, to understand the very fabric of physical reality, the quantum theory of gravity is simply the greatest scientific problem of our age. As it builds on two venerable theories, general relativity and quantum mechanics, both of which have reputations for messing with our heads, we shouldn't be too surprised to find that the concepts required to formulate a new theory of space and time stretch our comprehension to breaking point, and beyond. I hope that in this book I've at least been able to give you some sense for what is involved in both LQG and LQC, viewed through the prism of the scientific stories of Lee Smolin and Carlo Rovelli, two theorists who for thirty years have been closely involved in its inception, development, and evolution.

This Epilogue is an attempt to wind things up as neatly as possible, from both personal and scientific perspectives. It is based on a series of e-mail exchanges among Smolin, Rovelli, and myself and a joint Skype call on 15 December 2017.

Pursuing a quantum theory of gravity was always going to be a risky business, requiring great courage. Do you think this has paid off, on a personal level?

Committing to the study of quantum gravity requires nothing less than the abandonment of what might otherwise be regarded

Figure 31. Rovelli (left) and Smolin.

as a 'sensible' scientific career path. As young students, both Smolin and Rovelli were cautioned against making this choice by their teachers.

The nature of the career game was already clear to Smolin in his first term of graduate school at Harvard. Young, aspiring academic scientists are ranked in terms of their perceived ability to contribute to topics judged to be worthy, or trendy, with the winners getting faculty positions at the elite universities. Smolin had no interest in playing this game because it would have meant giving up on his very reason to do science. 'I once met [philosopher of science] Paul Feyerabend,' he explained, 'and he advised me to always do exactly what I want to do. He told me that if you have a clear and strong intention, nobody will put nearly as much energy into opposing you as you put into making your case.'

He well understood that such rebellious independence requires courage and might mean that his time in the Ivy League would be temporary. But, looking back on his career, his good fortune meant that his courage was not very much tested. He was instead rewarded with opportunities that allowed him to maintain his independence, opportunities which vastly exceeded the price to be paid in terms of doors that remained firmly closed. And, for the past 17 years, his position at the Perimeter Institute is, by his own account, 'the best job in the universe'.

'From a personal perspective it's been an extraordinary journey with much more satisfaction than I knew how to anticipate,' says Smolin. A lot of this satisfaction derives from the friendships that he has forged—with Rovelli and others—plus the immense satisfaction of seeing their ideas and results taken up by fabulously talented younger people, and watching their lives and careers thrive.

Rovelli feels much the same way. Earlier in his career as a theoretical physicist he would counsel young students who expressed a desire to follow the same path, pointing out the difficulties but encouraging them to take whatever path they liked. 'I am more careful today,' he says, 'but I still do so. We have only one life, so there is no point in being cautious. I advise students to study a lot before jumping into calculations, because I feel this has worked for me. And I also advise students not to follow advice, because this worked for me, too.'

This has been an extraordinary journey. 'It's been a fantastic trip, a real adventure,' says Rovelli. 'From when I first got to know Lee there have been moments of great enthusiasm, and moments of depression. But it's made me incredibly happy.' Smolin is reminded of a quote by Georges Braque, whose collaboration with Pablo Picasso pioneered cubism: 'The things that Picasso and I said to one another during those years will never be said again', said Braque, 'and even if they were, no one would understand them anymore. It was like being roped together on a mountain.'

'If I think about Carlo and his whole life in science from the time we met', says Smolin, 'I'm just incredibly impressed by what he's achieved and what he keeps provoking us to think about. I try to do the same, but I don't know if I succeed. The friendship and collaboration with Carlo changed my life. It made my life more complicated but richer, on both scientific and personal levels.'

For his part, Rovelli acknowledges his debt. 'Obviously, we couldn't have done what we've done together on our own. But my career in physics is due in large part to Lee. The way we work is much more like friendship, and this has been both a source of strength for me and a source of goodwill against the world. Irrespective of our disagreements, Lee has always been totally reliable and on my side. For me this is the best part of our story together.'

What about scientifically?

Rovelli's dream was to explore the subject and get some hints about physics at the Planck scale, and he believes this is what has happened. 'I feel that we've made some real progress,' says Rovelli. 'This might be major progress, or not. I really don't know. I'm perfectly happy with this situation but I would like to know more. I would like to live long enough to see some empirical confirmation of these hints, because at some point it will become clear. Whichever way it goes, it was definitely worthwhile.'

Smolin agrees. 'I'm also very happy with the structure of [loop] quantum gravity,' he says. 'What's marvelous is that if you think philosophically, as Carlo and I tend to do, then thinking about networks that are prior to geometry and from which geometry emerges is such a natural compromise of relationalism, background independence and locality.

'If I just wrote that down, I would feel quite justified in working with such a structure. The fact that it derives from an honest quantization of general relativity makes it very compelling. So whether the exact details of this operator or that operator turn out to be correct or not doesn't really concern me. What matters is that the general structure is extremely plausible: it's honestly derived and it's what you'd expect if you thought about it. I agree with Carlo—we don't know. This might be close to reality or reality might look nothing like this.'

In *Three Roads to Quantum Gravity*, Smolin famously predicted that we would have the basic framework of quantum gravity by 2010, or 2015 at the latest. Rovelli acknowledges that LQG is still incomplete in two fundamental respects. We lack direct empirical support for the theory, and there are numerous open questions that have yet to be resolved. 'The first is crucial', argues Rovelli. 'The second is in my opinion often over-emphasized. There are serious open issues in virtually *all* current theories. LQG, at least in some respects, is a possible theory of quantum spacetime. We do not yet know if it is right, but we do have one theory of quantum gravity. In this sense, I believe Lee's prediction was a good one.'

Smolin acknowledges Rovelli's generosity, but disagrees. He says: 'I believe I have to own up to the fact that when I wrote that I had in mind a scenario in which we would have experimental confirmation of the discreteness of quantum geometry, through gamma ray bursts and other astrophysical observations. Yes, the main predictions of LQG, such as the area and volume spectra, might turn out to be true, and that is personally very satisfying. But we don't know whether or not they are true—and that is equally frustrating.'

For the next generation of theorists, Smolin has this message: 'I tell them to regard us as the generation that failed to move physics definitively forward from how it was when it was handed to us in the 1970s and 1980s, and to freely salvage from our incomplete constructions whatever they find useful as they move beyond us to complete the revolution Einstein started.'

The most striking aspect of LQG is indeed its prediction that space itself is quantized. No other quantum theory of gravity predicts this. Would you say that this is your most satisfying achievement?

Smolin and Rovelli published their paper 'Discreteness of Area and Volume in Quantum Gravity' in May 1995 in the journal *Nuclear Physics B* (discussed at the start of Chapter 10). It is one of the most cited papers in the LQG community, and this result is the clearest evidence about the structure of space at the Planck scale. 'Achieving this result involved the proverbial mix of confusion, desperation and joy', explains Rovelli. 'The idea is actually Lee's, not mine, and Lee did the first calculation. We were in Verona, both distracted by romantic events, but I remember distinctively the hours in my apartment scribbling equations, and our astonishment and sheer joy when it started to really work. I think that that is the solid part of our achievement. If there is one thing that will always stay with me, it is that.'

Smolin agrees, but stresses that this was the culmination of a series of papers that discovered and established LQG, from the quantum geometries paper with Ted Jacobson, and the two founding papers of LQG with Rovelli. Smolin is also proud of his 1995 'linking' paper (see Chapter 11), which he regards as his best solo contribution to LQG: 'Even if not everyone recognizes it, this paper laid the foundation for all the work on horizons and entropy in LQG'.

It seems that the only real hope for gaining empirical support for LQG and LQC lie in future astronomical and cosmological observations. But will there ever be a consensus in favour of one approach to quantum gravity over the others?

To a certain extent, astrophysical observations have already given indications that disfavour approaches to quantum gravity that require certain Planck scale violations of special relativity, and those that require supersymmetry at the scale of physics probed at the LHC. 'Given the enormous past literature in these two directions,' says Rovelli, 'I would say that quantum gravity is now definitely confronting reality.'

The public perception of science is that this is all about 'truth', that there is always a right or wrong answer to a scientific question. It is true that we make progress, but what this progress reveals is that our understanding of what it true is necessarily contingent. 'A scientific discovery is very convincing,' says Rovelli, 'but at the same time scientific knowledge is never definitive and is always open to revision. This is a central point. Science establishes its credibility and its reliability through its refusal to be definitive and certain.' Something is true in science only until we discover that it's not, and we replace it with a new truth. For now.

As to where we should look, Smolin believes that the apparent invulnerability of quantum mechanics is vastly over-rated. 'As we push deeper and deeper into the quantum domain, I wouldn't be surprised if some anomalies surfaced which suggest corrections or completions of quantum mechanics.'

But, in science, experiments and observations are very rarely decisive. Studies of science history reveal a tendency for physicists to cling to cherished theories even in the face of sometimes overwhelming contrary evidence. More often there is a slow accumulation of indirect evidence which either supports or contradicts a theory, and which eventually results in its acceptance or rejection. Getting any kind of consensus would appear to be unlikely, even in the near-term.

Smolin is relaxed, however. 'In the absence of experimental confirmation, or a definitive argument, I don't think we need consensus,' he suggests. 'It would be premature. The world I would like to live in is one that rewards originality and independence, and encourages the development of diverse approaches towards unsolved problems.'

From recent experience it would seem that the main programmes in fundamental theoretical physics will require one, if not two, generations of theorists to work through to some conclusion. How does it feel to be working in an age in which you may never reach an answer?

Smolin recalls a summer school he attended in France as a graduate student. Richard Feynman delivered some of the lectures. Smolin had several conversations with Feynman, who in one of them remarked that he was very angry that he wasn't going to live long enough to see the answers to the questions that most interested him. 'I'm still in denial about this,' says Smolin, 'but it is troubling. I think the experimentalists should get to work, and use their imaginations to find some better experiments.

'It's not that experimentalists test theories—that's part of the story. It's often the case that divergences and disagreements *between* theories inspire experimentalists to invent tests that would not have been significant before. And I hope that's the case here. I think there must be a way out of this frustration, perhaps where we realize that something accessible today is already able to test the theories that we have.'

Rovelli feels differently. 'Look, Copernicus died without any strong evidence that the Earth goes around the Sun. Boltzmann died without any evidence that atoms and molecules are the sources of heat and entropy. So many great novelists died before their novels were even published. This is fine—as theorists we still have our jobs to do.

'Of course I'm curious. Of course I would like to know. But I'm not pessimistic. It's common to think that nothing is happening but then you look back and realize that things *are* happening. Today the attitude towards supersymmetry is changed compared to what it was even five years ago. We now have experimental confirmation of gravitational waves, and soon we will have the Event Horizon Telescope.* I wouldn't be surprised if we get empirical information that we're not even thinking about right now.'

The theoretical physics community remains polarized, but do you think there are prospects for a future reconciliation of strings and loops?

Surprisingly, given his perceived role in the 'String Wars' of 2006, Smolin thinks that that the only episode in his career that demanded real courage occurred in the late 1990s, when he chose to work on unifying string theory and LQG. This led to a degree of isolation, and drew some criticism from the community to which he had belonged:

> These were people I had enormous respect and affection for and it was impossible to convince them of my view that, from the beginning, string theory and LQG were like conjoined twins in that they express the same physical idea of the duality between quantum fields and extended objects.

* The Event Horizon Telescope (EHT) is a project to create an Earth-sized virtual telescope from a global network of eight radio telescopes, using a technique called very long baseline interferometry. The EHT is being used to study the Milky Way's central supermassive black hole (Sagittarius A*) and the black hole at the centre of the elliptical galaxy Messier 87, with an angular resolution similar to the black hole event horizon. Disks carrying data from the South Pole Telescope were delivered to the MIT Haystack Observatory on 13 December 2017, and data analysis was begun early in 2018.

As I hope *Quantum Space* shows, LQG and string theory are in some respects quite closely related and, whilst Smolin's efforts to bring them together have so far been frustrated, the possibility of some kind of reconciliation appears to be growing. He remains hopeful, but he believes that the string theory community has since moved on. Today, few string theorists seem to be concerned with strings per se: they're more interested in the AdS/CFT correspondence and, very recently, in an intriguing set of ideas concerned with the emergence of spacetime from the entanglement of quantum information. 'What is really exciting to me now is the fact that LQG offers an ideal tool kit to express holography and quantum entanglement', Smolin says. 'In fact, Roger Penrose's original motivation for spin networks was to use quantum entanglement to express Mach's principle in a relational context—it's there in his first handwritten notes on spin networks. Penrose's original spin geometry theorem is an early anticipation of the idea of space arising from entanglement. There is much scope for expressing these ideas in the language of LQG.'

There are promising signs. At a recent (2017) LQG conference in Warsaw, Herman Verlinde, a leading string theorist at Princeton University (and identical twin brother of Erik), talked about possible relationships between strings and loops, and the project remains very much alive.

'I have always been more sceptical,' says Rovelli, 'but I am changing my mind.'

GLOSSARY

acceleration. The rate of change of velocity (speed) with time, usually given the symbol a, as in $F = ma$, or force equals mass times acceleration (Newton's second law of motion).

ADM formalism. A greatly simplified elaboration of the constrained Hamiltonian formulation of general relativity, named for its authors Richard Arnowitt, Stanley Deser, and Charles Misner, and first published in 1959. This formulation plays a significant role in the canonical approach to quantum gravity and represents an important step towards the Wheeler–DeWitt equation.

angular momentum. The rotational equivalent of the more familiar linear momentum. In classical mechanics, angular momentum is a vector (or pseudo-vector) quantity derived from an object's rotational inertia and rotational or angular velocity. In quantum mechanics there are two types of angular momentum associated with orbital motion (for example, of an electron in an atom) and spin. These may combine to give a total angular momentum. In both classical and quantum mechanics, angular momentum is a conserved quantity.

antiparticle. Identical in mass to an 'ordinary' particle but of opposite charge. For example, the antiparticle of the electron (e^-) is the positron (e^+). The antiparticle of a red quark is an anti-red antiquark. Every particle in the standard model has an antiparticle. It is hypothesized that neutrinos and antineutrinos (which differ only in terms of their chirality—neutrinos are 'left-handed') may actually be two different states of the same particle. These are called Majorana particles for Italian theorist Ettore Majorana. Experiments are underway to find out if this is indeed the case.

Ashtekar variables. The 'new variables' introduced by Abhay Ashtekar in 1986 allowed a constrained Hamiltonian form of general relativity to be formulated in terms of spin connections. This was a fundamentally important step on the way to developing loop quantum gravity.

atom. From the Greek *atomos*, meaning indivisible or uncuttable. Originally intended to denote the ultimate constituents of matter, the word atom now signifies the fundamental constituents of individual chemical elements. Thus, water consists of molecules of H_2O, which is composed of two atoms of hydrogen and one atom of oxygen. The atoms in turn consist of protons and neutrons, which are bound together to form a central nucleus, and electrons whose wavefunctions form characteristic patterns called orbitals around the nucleus.

atto. A prefix denoting a billion billionth (10^{-18}). An attometre (am) is 10^{-18} metres, or a thousandth of a femtometre. The radius of a proton is about 850 am. The LIGO gravitational wave observatory is sensitive to displacements on the order of 1 am.

bare mass. The hypothetical mass that a particle would possess if it could be separated from the quantum fields which it generates or with which it interacts. The observed mass of the particle is then the bare mass plus mass generated by interactions with the quantum fields.

baryon. From the Greek *barys*, meaning heavy. Baryons form a subclass of hadrons. They are heavier particles which experience the strong nuclear force and include the proton and neutron. They are composed of triplets of quarks.

Bell's theorem/inequality. Devised by John Bell in 1966. The simplest extension of quantum theory which resolves the problem of the collapse of the wavefunction and the 'spooky action-at-a-distance' that this seems to imply involves the introduction of local hidden variables which govern the properties and behaviour of quantum particles. Bell's theorem states that the predictions of any local hidden variable theory will not always agree with the predictions of quantum theory. This is summarized in Bell's inequality—the predictions of local hidden variable theories cannot exceed a certain maximum limit. But quantum theory predicts results that for certain experimental arrangements *will* exceed this limit. Bell's inequality therefore allows a direct experimental test.

beta particle. A high-speed electron emitted from the nucleus of an atom undergoing beta radioactive decay. *See* beta radioactivity/decay.

beta radioactivity/decay. First discovered by French physicist Henri Becquerel in 1896 and so named by Ernest Rutherford in 1899. An example of a weak force decay, it involves transformation of a down quark in a neutron into an up quark, turning the neutron into a proton with the emission of a W⁻ particle. The W⁻ decays into a high-speed electron (the 'beta particle') and an electron antineutrino.

Big Bang. Term used to describe the cosmic 'explosion' of spacetime and matter during the early moments in the creation of the universe, about 13.8 billion years ago. Originally coined by maverick physicist Fred Hoyle as a derogatory term, overwhelming evidence for a Big Bang 'origin' of the universe has since been obtained through the detection and mapping of the cosmic microwave background radiation, the cold remnant of hot radiation thought to have disengaged from matter about 380,000 years after the Big Bang.

big bounce. The current inflationary Big Bang model (sometimes called the Λ-CDM model) is based on classical general relativity and predicts that the universe originated in a Big Bang 'singularity' involving infinite density and temperature at 'time zero'. In contrast, in loop quantum cosmology there can be no volume of space smaller than the elementary quantum of volume, and computer simulations suggest that as the universe contracts towards this smallest volume, at a critical density much less than the Planck density, the universe will 'bounce' and start to expand.

billion. One thousand million, 10^9 or 1,000,000,000.

black hole. A name popularized (though not, as many suggest, coined) by John Wheeler. A black hole is a region of spacetime containing so much mass-energy that its escape velocity—the speed required to escape its gravitation pull—is greater than the speed of light. This idea actually dates back to the eighteenth century, but came to prominence through the work of Karl Schwarzschild, who in 1916 was the first to derive formal solutions for Einstein's gravitational field equations. *See also* Schwarzschild solution/radius.

Bohr radius. The orbital distance of the electron as measured from the proton in a hydrogen atom. In Bohr's model of the atom, published in 1913, Bohr calculated this distance from a collection of fundamental physical constants (including Planck's constant, the speed of light, and the mass and charge of the electron). In Schrödinger's wave mechanics, the electron is spherically 'distributed' within its lowest-energy orbital, but has the highest probability for being found at a distance corresponding to the Bohr radius, a little over 0.0529 nanometres.

boson. Named for Indian physicist Satyendra Nath Bose. Bosons are characterized by integral spin quantum numbers (1, 2, ..., etc.) and, as such, are not subject to Pauli's exclusion principle. Bosons are involved in the transmission of forces between matter particles and include the photon (electromagnetism), the W and Z particles (weak force), and gluons (colour force). Particles with spin 0 are also called bosons but these are not involved in transmitting forces. Examples include the pions and the Higgs boson. The graviton, the hypothetical particle of the gravitational field, is believed to be a boson with spin 2.

bottom quark. Also sometimes referred to as the 'beauty' quark. A third-generation quark with charge -⅓, spin ½ (fermion), and a 'bare mass' of $4.18 \text{ GeV}/c^2$. It was discovered at Fermilab in 1977, through the observation of the upsilon, a meson formed from bottom and antibottom quarks.

Calabi–Yau space. A particular type of space (or manifold, or 'compactification') which has properties suited to the representation of the six extra hidden spatial dimensions required by superstring theories. The holes in the Calabi–Yau space accommodate low-energy string vibrations, so the number of holes determines the number of families of particles that will result.

canonical (approach to quantum gravity). The canonical approach to the formulation of a quantum field theory was devised by Paul Dirac, as a 'method of classical analogy'.* It starts with the Hamiltonian form of the classical field theory and replaces classical variables (such as linear or angular momentum) with their equivalent quantum-mechanical operators, in ways that preserve the formal structure of the classical theory. The canonical approach to quantum gravity then starts with a constrained Hamiltonian form of general relativity and involves finding a way to 'quantize' space.

CERN. Acronym for Conseil Européen pour la Recherche Nucléaire (the European Council for Nuclear Research), founded in 1954. This was renamed the Organisation Européenne pour la Recherche Nucléaire

---

* P. A. M. Dirac, *The Principles of Quantum Mechanics*, 4th edn, Oxford University Press, Oxford, 1958, p. 84.

(European Organization for Nuclear Research) when the provisional Council was dissolved, but the acronym CERN was retained. CERN is located in the northwest suburbs of Geneva near the Swiss–French border.

charm quark. A second-generation quark with charge $+\frac{2}{3}$, spin $\frac{1}{2}$ (fermion), and a 'bare mass' of $1.28\,\text{GeV}/c^2$. It was discovered simultaneously at Brookhaven National Laboratory and SLAC in the 'November revolution' of 1974 through the observation of the J/ψ, a meson formed from a charm and anticharm quark.

classical mechanics. The system of mechanics embodied in Newton's laws of motion and the law of universal gravitation, although the study of mechanics predates Newton. The system deals with the influence of forces on the motions of larger, macroscopic bodies at speeds substantially less than light. Although 'classical', the system remains perfectly valid today within the limits of its applicability.

cold dark matter. A key component of the current Λ-CDM model of inflationary Big Bang cosmology, thought to account for about 26.0 per cent of the mass-energy of the universe. The constitution of cold dark matter is unknown, but is thought to consist largely of 'non-baryonic' matter, i.e. matter that does not involve protons or neutrons, most likely particles not currently known to the standard model.

collapse of the wavefunction. In most quantum systems, the wavefunction of a quantum entity will be delocalized over a region of space (the quantum entity may be here, there, or everywhere within the boundary of the wavefunction), yet when a measurement is made the result is localized to a specific position (the entity is here). Similarly, in a quantum measurement in which a number of different outcomes is possible (such as spin-up or spin-down), it is necessary to form a superposition of the wavefunctions describing these outcomes. The probability of getting a specific result is then related to the square of the amplitude of the corresponding wavefunction in the superposition. In either case, the wavefunction or superposition is said to 'collapse'. A number of possible outcomes converts to one outcome only, and all the other possibilities disappear.

colour charge. A property possessed by quarks in addition to flavour (up, down, strange, charm, top, bottom). Unlike electric charge, which comes in two varieties—positive and negative—colour charge comes in three varieties, which physicists have chosen to call red, green, and blue. Obviously, the use of these names does not imply that quarks are 'coloured' in the conventional sense. The colour force between quarks is carried by coloured gluons.

colour force. The strong force responsible for binding quarks and gluons together inside hadrons. Unlike more familiar forces, such as gravity and electromagnetism which get weaker with increasing distance, the colour force acts like a piece of elastic or spring, tethering the quarks together. When the quarks are close together, the elastic or spring is relaxed and the quarks behave as though they are entirely free. But as the quarks are pulled apart, the elastic or spring tightens and keeps the quarks 'confined'. The strong nuclear force which binds protons and neutrons together inside atomic nuclei results from a kind of 'leakage' of the colour force beyond the boundaries of the nucleons.

complementarity. The principle of complementarity was devised by Niels Bohr and is a key pillar of the Copenhagen interpretation of quantum mechanics. According to this principle, quantum wave-particles will exhibit wave-like and particle-like behaviour in mutually exclusive experimental arrangements, but it is impossible to devise an arrangement that will show both types of behaviour simultaneously. Such behaviour is, however, not contradictory; it is complementary.

complex number. A complex number is formed by multiplying a real number by the square root of -1, written i. The square of a complex number is then a negative number: for example, the square of $5i$—$(5i)^2$—is -25. Complex numbers are used widely in mathematics to solve problems that are impossible using real numbers only.

conservation law. A physical law which states that a specific measureable property of an isolated system does not change as the system evolves in time. Measureable properties for which conservation laws have been established include mass-energy, linear and angular momentum, electric and colour charge, isospin, etc. According to a theorem by German mathematician Amalie Emmy Noether, each conservation law can be traced to a specific continuous symmetry of the system.

Copenhagen interpretation. Developed by Niels Bohr, Werner Heisenberg, and Wolfgang Pauli as a way of thinking about the nature of elementary quantum wave-particles as described by quantum mechanics. Depending on the experimental setup, such wave-particles will exhibit wave-like or particle-like behaviour. But these behaviours are complementary: in this kind of experiment the wave-particle is a wave, in this other kind of experiment it is a particle, and it is meaningless to ask what the wave-particle *really is*.

cosmic background radiation. Some 380,000 years after the Big Bang, the universe had expanded and cooled sufficiently to allow hydrogen nuclei (protons) and helium nuclei (consisting of two protons and two neutrons) to recombine with electrons to form neutral hydrogen and helium atoms. At this point, the universe became 'transparent' to the residual hot radiation. Further expansion shifted and cooled this hot radiation to the microwave and infrared regions with a temperature of just 2.7 kelvin (−270.5°C), a few degrees above absolute zero. This microwave background radiation was predicted by several theorists and was discovered accidentally by Arno Penzias and Robert Wilson in 1964. The COBE, WMAP, and Planck satellites have since studied this radiation in great detail.

cosmic inflation. A rapid exponential expansion of the universe thought to have occurred between 10^{-36} and 10^{-32} seconds after the Big Bang. Discovered by American physicist Alan Guth in 1980, inflation helps to explain the large-scale structure of the universe that we observe today. *See also* slow-roll inflation.

cosmic scale factor. A dimensionless quantity usually given the symbol a. If d is the proper distance between two points at some time t, and d_0 is this same distance at some earlier fixed time t_0, then $a = d/d_0$. In a universe which expands at a constant rate, d is greater than d_0 and a is fixed and greater than 1. The Hubble constant H is defined as the ratio of $\dot{a}$, the rate of change of a with time, divided by a. We can think of $\dot{a}$ as the velocity of expansion. Because d_0 is fixed, it doesn't change with time, so $\dot{a} = \dot{d}/d_0$, where $\dot{d}$ is the rate of change of proper distance with time. We can replace the rate of change of distance $\dot{d}$ with the *recession velocity*, v. Hence, $H = \dot{a}/a = \dot{d}/d = v/d$, or $v = Hd$ —the speed of recession is directly proportional to distance, which is just Hubble's law. However, we now know that the rate of expansion is *not* constant—the velocity $\dot{a}$ is itself increasing with time, or the expansion is *accelerating*. This means that the Hubble constant varies over time. *See also* Hubble's law.

cosmological constant. Albert Einstein initially resisted the idea that the universe is dynamic—that it could expand or contract—and fudged his equations to produce static solutions. Concerned that conventional gravity would be expected to overwhelm the matter in the universe and cause it to collapse in on itself, Einstein introduced a 'cosmological constant'—a kind of negative or repulsive form of gravity—to counteract the effect. When evidence accumulated that the universe is actually expanding, Einstein regretted his action, calling it the biggest blunder he had ever made in his life. But, in fact, further discoveries in 1998 suggested that the expansion of the universe is actually accelerating. When combined with satellite measurements of the cosmic microwave background radiation these results have led to the suggestion that the universe is pervaded by 'dark energy', accounting for about 69.1 per cent of the mass-energy of the universe. One form of dark energy requires the reintroduction of Einstein's cosmological constant.

cosmological redshift. Light emitted from a distant galaxy or supernova will consist of a range of frequencies or wavelengths corresponding to emissions from different energy states of a variety of atoms. This light travels to Earth at the speed of light. But in the time it takes for this light to reach Earth, the space in between has expanded, causing the light to appear red-shifted (lower frequency or longer wavelength). Just as with the Doppler Effect, the extent of the redshift can be used to determine the apparent speed with which the distant galaxy or supernova is receding from Earth due to the expansion of the universe. A graph of recession speed versus distance appears linear—the speeds appear directly proportional to the distance (Hubble's law). In fact, measurements on a certain class of supernovae in 1998 demonstrated that the expansion of the universe is actually accelerating. *See also* Hubble's law.

covariant (approach to quantum gravity). Another approach to a quantum theory of gravity starts with a quantum field theory and seeks to find ways to make this conform to Einstein's principle of general covariance. In other words, start with a quantum field theory which presupposes a background spacetime and seek to make it background-independent.

dark energy. *See* cosmological constant.

dark matter. Discovered in 1934 by Swiss astronomer Fritz Zwicky as an anomaly in the measured mass of galaxies in the Coma cluster (located in the constellation Coma Berenices). Zwicky observed that the rotation speeds of galaxies near the cluster edge are much faster than predicted from the mass of all the observable galaxies, implying that the mass of the cluster is actually much larger. As much as 90 per cent of the mass required to explain the rotation curve appeared to be 'missing', or invisible. This missing matter was called 'dark matter'. Subsequent studies favour a form of dark matter called 'cold dark matter'. *See* cold dark matter.

de Broglie relation. Deduced by Louis de Broglie in 1923, this equation relates a wave-like property (the wavelength, λ) of a quantum wave-particle to a particle-like property (linear momentum, p). The relation is $\lambda = h/p$, where h is Planck's constant. For 'everyday' objects of macroscopic size (such as a tennis ball), the wavelength predicted by the de Broglie relation is much too short to observe. But microscopic entities such as electrons have measureable wavelengths, typically 100,000 times shorter than visible light. A beam of electrons can be diffracted and will show two-slit interference effects. Electron microscopes are used routinely to study the structures of inorganic and biological samples.

diffraction. A physical phenomenon which is revealed when light is forced to squeeze through a narrow aperture or circular hole that is comparable in size to the wavelength of the light. It is most easily interpreted in terms of a wave description. In this description, every point on the wavefront acts as a source for a new spherical wave. If undisturbed, these spherical waves combine to form a new wavefront and in this way the wave moves forwards through space. But when forced through a narrow aperture or hole, the wavefront 'bends' and the spherical waves may no longer combine quite so smoothly. Such combination depends not only on the amplitude (height) of the wave but also is phase (the point in its peak-to-trough cycle). The result is interference—a diffraction pattern consisting of bright and dark fringes.

dressed mass. The mass derived from a quantum wave-particle's self-energy, the result of interactions with the system from which it is physically inseparable. For example, an electron acquires self-energy by interacting with its own self-generated electromagnetic field.

electric charge. A property possessed by quarks and leptons (and, more familiarly, protons and electrons). Electric charge comes in two varieties—positive and negative—and the flow of electrical charge is the basis for electricity and the power industry.

electromagnetic force. Electricity and magnetism were recognized to be components of a single, fundamental force through the work of several experimental and theoretical physicists, most notably English physicist Michael Faraday and Scottish theoretician James Clerk Maxwell. The electromagnetic force is responsible for binding electrons with their nuclei inside atoms, and binding atoms together to form the great variety of molecular substances.

electron. Discovered in 1897 by English physicist Joseph John Thomson. The electron is a first-generation lepton with a charge −1, spin ½ (fermion), and mass 0.51 MeV/c^2.

electroweak force. Despite the great difference in scale between the electromagnetic and weak nuclear forces, these are facets of what was once a unified electroweak force, thought to prevail during the 'electroweak epoch', between 10^{-36} and 10^{-12} seconds after the Big Bang. The combination of electromagnetic and weak nuclear forces in a unified field theory was first achieved by Steven Weinberg and independently by Abdus Salam in 1967–8.

empiricism. One of several philosophical perspectives on the acquisition of human knowledge. In an ardently empiricist philosophy, knowledge is firmly linked to experience and evidence—'seeing is believing'. If we can't directly experience or acquire even indirect evidence for the existence of an entity, then we have no grounds for believing that it really exists. Such an entity would then be regarded as metaphysical (leading the Austrian physicist and arch-empiricist Ernst Mach famously to reject the reality of atoms). Today few scientists are likely to dispute the existence of an independent reality (the Moon is still there even when nobody is looking), and few will dispute the existence of entities such as electrons and quarks which can be 'observed' only indirectly. But it is possible that some scientists will adopt an empiricist or antirealist interpretation of the concepts used in the theoretical *representation* of such entities. The Copenhagen interpretation is an example of an antirealist perspective on the quantum theory formalism. *See also* instrumentalism.

equivalence principle. In Einstein's 'happiest thought' he realized that in free fall we feel neither acceleration nor gravity: our experience of acceleration is equivalent to our experience of gravity. This implies that the inertial mass of an object (a measure of the object's resistance to acceleration, or the m in $F = ma$) is the same as its gravitational mass (the mass responsible for exerting the force of gravity, or the m_1 or m_2 in $F = Gm_1m_2/r^2$).

Euclidean space. Named for the ancient Greek mathematician Euclid of Alexandria. This is the familiar geometry of 'ordinary' three-dimensional or 'flat' space, typically described in terms of Cartesian coordinates (x,y,z) and in which the angles of a triangle add up to 180°, the circumference of a circle is given by 2π times its radius, and parallel lines never meet.

exclusion principle. *See* Pauli exclusion principle.

femto. A prefix denoting a million billionth (10^{-15}). A femtometre (fm) is 10^{-15} metres, a thousand attometres, or a thousandth of a picometre. The radius of a proton is about 0.85 fm.

fermion. Named for Italian physicist Enrico Fermi. Fermions are characterized by half-integral spins ($\frac{1}{2}$, $\frac{3}{2}$, etc.) and include quarks and leptons and many composite particles produced from various combinations of quarks, such as baryons.

flat space. *See* Euclidean space.

flavour. A property which distinguishes one type of quark from another in addition to colour charge. There are six flavours of quark which form three generations up, charm, and top with electric charge $+\frac{2}{3}$, spin $\frac{1}{2}$, and masses of 1.8–3.0 MeV/c^2, 1.28 GeV/c^2, and 173 GeV/c^2, respectively, and down, strange, and bottom with electric charge $-\frac{1}{3}$, spin $\frac{1}{2}$, and masses 4.5–5.3 MeV/c^2, 95 MeV/c^2, and 4.18 GeV/c^2, respectively. The term flavour is also sometimes applied to leptons, with the electron, muon, tau, and their corresponding neutrinos distinguished by their 'lepton flavour'. *See* lepton.

force. Any action that changes the motion of an object. In Isaac Newton's three laws of motion forces are 'impressed', meaning that the actions involve some kind of physical contact between the object and whatever is generating the force (such as another object). The exception to the rule is Newton's force of gravity, which appears to act instantaneously and at a distance (no obvious contact between gravitating objects, such as the Earth and the Moon). This problem is resolved in Einstein's general theory of relativity.

frame-dragging. An object with mass, such as the Earth, moving or rotating in a stationary or steady-state gravitational field, is predicted by general relativity to drag spacetime around with it. This is an effect first deduced by Austrian physicists Josef Lense and Hans Thirring and is sometimes referred to as the Lense–Thirring effect. It is a kind of gravitational analogue of electromagnetic induction. Frame-dragging was observed by Gravity Probe B, but problems with the onboard gyroscopes meant that although the extent of precession of the gyroscopes was entirely consistent with the prediction of general relativity, there was considerable uncertainty in the measured value.

frame of reference. Defined by a coordinate system against which physical points can be located and measurements can be referred. The frame may refer to a static coordinate system in which motion is measured, or it may refer to the type of motion involved (as in a rotating frame of reference). In special relativity, particular attention is given to *inertial frames of reference*, in which objects experience no net force and so continue in a state of rest (rest frame) or motion at constant velocity in a straight line (Newton's first law of motion).

general covariance. If there is no absolute frame of reference or absolute system of coordinates, then the forms of the laws of physics must remain the same (they must be invariant) on any change of coordinate system.

general (theory of) relativity. Developed by Einstein in 1915. The general theory of relativity incorporates special relativity and Newton's law of universal gravitation in a geometric theory of gravitation. Einstein replaced the 'action-at-a-distance' implied in Newton's theory with the movement of massive bodies in a curved spacetime. In general relativity, matter tells spacetime how to curve, and the curved spacetime tells matter how to move.

g-factor. A constant of proportionality between the (quantized) angular momentum of an elementary or composite particle and its magnetic moment, the direction the particle will adopt in a magnetic field. There are actually three g-factors for the electron, one associated with its spin, one associated with the angular momentum of the electron orbital motion in an atom, and one associated with the sum of spin and orbital angular momentum. Dirac's relativistic quantum theory of the electron predicted a g-factor for electron spin of 2. The value recommended in 2010 by the International Council for Science Committee on Data for Science and Technology (CODATA) task group is 2.00231930436153. The difference is due to quantum electrodynamic effects.

giga. A prefix denoting billion. A giga electron volt (GeV) is a billion electron volts, 10^9 eV or 1,000 MeV.

gluon. The carrier of the strong colour force between quarks. Quantum chromodynamics requires eight, massless colour force gluons which themselves carry colour charge. Consequently, the gluons participate in the force rather than simply transmit it from one particle to another. Ninety-nine per cent of the mass of protons and neutrons is thought to be energy carried by gluons and quark–antiquark pairs created by the colour field.

grand unified theory (GUT). Any theory which attempts to unify the electromagnetic, weak, and strong nuclear forces in a single structure is an example of a grand unified theory. The first example of a GUT was developed by Sheldon Glashow and Howard Georgi in 1974. GUTs do not seek to accommodate gravity: theories that do are generally referred to as theories of everything (TOEs).

gravitational force. The force of attraction experienced between all mass-energy. Gravity is extremely weak and has no real part to play in the interactions between atoms and subatomic and elementary particles which are rather governed by the colour force, weak force, and electromagnetism. The effects of gravity are described by Einstein's general theory of relativity and approximated in Newton's theory of universal gravitation.

graviton. A hypothetical particle which carries the gravitational force in covariant quantum gravity, or a quasiparticle which appears in canonical quantum gravity. If it exists, the graviton would be a massless, chargeless boson with a spin quantum number of 2. It is the particle analogue of gravitational waves. However, it is very unlikely that a graviton will ever be detected.

hadron. From the Greek *hadros*, meaning thick or heavy. Hadrons form a class of particles which experience the strong nuclear force and are therefore composed of various combinations of quarks. This class includes baryons, which are composed of three quarks, and mesons, which are composed of one quark and an antiquark.

Hamiltonian (mechanics/function/operator). Hamiltonian mechanics is a reformulation of classical Newtonian mechanics that was established by Irish physicist William Rowan Hamilton in 1833. The classical Hamiltonian function is the sum of the system's kinetic and potential energies. The kinetic energy depends on the momentum and is independent of the spatial coordinates used to describe the system. The potential energy depends on the spatial coordinates and is independent of the momentum. In quantum mechanics (in the form of the Schrödinger equation), the classical momentum variable is replaced by its equivalent quantum mechanical operator, and the classical Hamiltonian function becomes the Hamiltonian operator for the total energy of the system.

Heisenberg's uncertainty principle *see* uncertainty principle.

hidden variables. The simplest way to modify or extend conventional quantum mechanics to eliminate the collapse of the wavefunction is to introduce hidden variables. Such variables govern the properties and behaviour of quantum wave-particles but by definition cannot be observed directly. If the resulting extension is required to ensure that individual quantum entities possess specific properties at all times (in other words, the entities are 'locally real'), then the hidden variables are said to be local. If the extension is required to ensure that quantum entities possess specific properties in a collective sense, then the hidden variables may be non-local.

Higgs boson. Named for English physicist Peter Higgs. All Higgs fields have characteristic field particles called Higgs bosons. The term 'Higgs boson' is typically reserved for the electroweak Higgs, the particle of the Higgs field first used in 1967–8 by Steven Weinberg and Abdus Salam to account for electroweak symmetry-breaking. The electroweak Higgs boson was discovered at CERN's Large Hadron Collider, a discovery announced on 4 July 2012. It is a neutral, spin-0 particle with a mass of about 125 GeV/c^2.

Higgs field. Named for English physicist Peter Higgs. A generic term used for any background quantum field added to a field theory to trigger symmetry-breaking through the Higgs mechanism. The existence of the Higgs field used to break the symmetry in a quantum field theory of the electroweak force is strongly supported by the discovery of the Higgs boson at CERN.

Higgs mechanism. Named for English physicist Peter Higgs, but also often referred to using the names of other physicists who independently discovered the mechanism in 1964: Robert Brout, François Englert, Higgs, Gerald Guralnik, Carl Hagen, and Tom Kibble. The mechanism describes how a background quantum field—called the Higgs field—can be added to a field theory to break a symmetry. In 1967–8 Steven Weinberg and Abdus Salam independently used the mechanism to develop a field theory of the electroweak force.

Hubble's law. The observation, first reported by American astronomer Edwin Hubble, that distant galaxies recede from us at speeds directly proportional to their distances. This relationship is summarized by the equation $v = Hd$, in which v is the recession speed of the galaxy, d is its distance from Earth, and H is the Hubble parameter, which has a value of 67.7 kilometres per second per megaparsec (based on analysis of Planck satellite mission data reported in 2015). Studies of certain types of distant supernovae in 1998 revealed that the expansion of the universe is actually accelerating, so the Hubble parameter is not a constant. *See also* cosmic scale factor, cosmological redshift.

inertial frame of reference. *See* frame of reference.

inflation. *See* cosmic inflation.

inflaton (field). Rapid inflation in the very earliest moments following the Big Bang is thought to have been responsible for 'flattening' the universe, and so resolving the flatness and horizon problems (and another problem relating to magnetic monopoles). One version of the inflationary scenario is based on the notion of an inflaton field. This is a scalar quantum field which acts much like a Higgs field (and there are some suggestions that the inflaton field and the Higgs field are one and the same). The particle of the inflaton field is called the inflaton. It works something like this. The 'empty' vacuum state of the inflaton field is a so-called 'false vacuum'—it has a non-zero energy which acts like a cosmological constant and is responsible for the rapid expansion of spacetime. After a brief burst of inflation, the false vacuum decays to the 'true' vacuum. Spacetime continues to expand but now much more gently, with a little residual dark energy responsible for the acceleration of this expansion. *See also* cosmic inflation, slow-roll inflation.

information paradox. Because it emits Hawking radiation, the surface area of a black hole is expected to shrink in size and eventually 'evaporate'. This is a problem because there is a connection between the entropy of a black hole and the quantum information stored on its surface. If this information is lost or irretrievably scrambled by the Hawking radiation and then completely disappears from the universe, this violates an important principle of quantum mechanics: all the information coded in the wavefunction must be preserved, in a kind of conservation of probability. The irretrievable loss of information implied by black hole evaporation then directly contradicts the very basis and structure of quantum theory.

instrumentalism. An extreme form of empiricism at the level of theoretical representation. Theoretical concepts and structures are judged to serve simply as instruments or tools that we use to summarize past experience and predict the future, and have no real meaning beyond this. Theories are simply a means to an end. *See also* empiricism.

interference (waves). Like diffraction, interference is best interpreted in terms of a wave model. When waves from two different sources (or waves from one source that are passed through two adjacent holes or slits) combine together, the result depends on both the amplitudes of the waves and their relative phases (the positions in their peak-to-trough cycles). The waves may combine constructively, producing a stronger wave, or destructively, producing a weaker wave or no wave at all. Such interference is typically manifested by alternating bright and dark fringes. *See also* diffraction.

kaon. A group of spin-0 mesons consisting of up, down, and strange quarks and their antiquarks. These are K^+ (up-anti-strange), K^- (strange-anti-up), K^0 (mixtures of down-anti-strange and strange-anti-down) with masses 493.7 MeV/c^2 ($K^\pm$) and 497.6 MeV/c^2 (K^0).

Λ-CDM. An abbreviation of lambda-cold dark matter. Also known as the 'concordance model' or the 'standard model' of Big Bang cosmology. The

Λ-CDM model accounts for the large-scale structure of the universe, the cosmic microwave background radiation, the accelerating expansion of the universe, and the distribution of elements such as hydrogen, helium, lithium, and oxygen. Based on the most recent Planck satellite data, the model is consistent with a universe that 'began' 13.8 billion years ago and in which 69.1 per cent of its mass-energy is in the form of dark energy (reflected in the size of the cosmological constant, Λ); 26.0 per cent is cold dark matter; leaving the visible universe—galaxies, stars, planets, gas, and dust—to account for just 4.9 per cent.

Leggett inequality. Named for English physicist Anthony Leggett as an extension of the logic of Bell's theorem and Bell's inequality. The introduction of local hidden variables implies two logical consequences: measurements involving entangled pairs are not affected by the way the experimental apparatus is *set up* and they are not affected by the *results* of any measurement that is actually made on one, the other, or both particles in the pair. Leggett defined a class of 'crypto' non-local hidden variable theories in which the experimental setup can indeed affect the outcome, but the actual results cannot. Such theories do not predict all the possible results that quantum theory predicts and Leggett was able to develop an inequality that could be subjected to a direct test.

lepton. From the Greek *leptos*, meaning small. Leptons form a class of particles which do not experience the strong nuclear force and combine with quarks to form matter. Like quarks, leptons form three generations, including the electron, muon, and tau with electric charge −1, spin ½, and masses 0.51 MeV/c^2, 105.7 MeV/c^2, and 1.78 GeV/c^2, respectively, and their corresponding neutrinos. The electron, muon, and tau neutrinos carry no electric charge, have spin ½, and are believed to possess very small masses (necessary to explain the phenomenon of neutrino oscillation, the quantum-mechanical mixing of neutrino flavours such that the flavour may change over time).

LHC. Acronym for Large Hadron Collider. The world's highest-energy particle collider, designed to produce proton–proton collision energies of 14 TeV. The LHC is 27 kilometres in circumference and lies 175 metres beneath the Swiss–French border at CERN near Geneva. The LHC, operating at proton–proton collision energies of 7 and 8 TeV, produced evidence which led to the discovery of the Higgs boson in July 2012. After a two-year shutdown, it began operations in 2015 at a collision energy of 13 TeV.

loop quantum cosmology (LQC). A form of quantum cosmology based on the principles of loop quantum gravity. In LQC, the Big Bang singularity is avoided because space cannot be compressed to a volume smaller than a single quantum of the volume of space. LQC also predicts superinflation, but this is insufficient to explain the large-scale structure of our universe, so cosmic inflation via an inflaton field is still required. *See also* superinflation, spinfoam.

loop quantum gravity (LQG). A quantum theory of gravity derived from the canonical approach based on a reformulation of the constrained Hamiltonian form of general relativity using spin connections (Ashtekar variables). LQG predicts the existence of quanta of area and volume of space on the order of the square and cube of the Planck length, respectively. *See also* Planck scale, spinfoam.

Mach's principle. Summarizes the notion that all motion—inertial and accelerating—must be relative. Newton's bucket argument for the absolute nature of rotational motion fails because this is still rotation relative to the rest of the universe and we can't tell if it is the bucket that is spinning in a stationary universe or the universe is spinning around a stationary bucket. Local physical laws are determined by the large-scale structure of the universe. The principle is actually credited to Einstein and was a significant inspiration for his development of general relativity. Its status is today still very much disputed, but frame-dragging suggests a mechanism by which distant objects may exert local influences.

mass. In classical mechanics, the mass of a physical object is a measure of its resistance to changes in its state of motion under the influence of an applied force, assumed to be related to the 'quantity of matter' that it contains. As such, it is a 'primary' quality of material substance. In special relativity and quantum physics, our understanding of the nature of mass changes quite dramatically. Mass becomes a measure of the energy content of an object ($m = E/c^2$), and the mass of elementary particles is traced to the energies associated with different kinds of quantum fields.

mass renormalization. *See* renormalization.

mega. A prefix denoting million. A mega electron volts (MeV) is a million electron volts, 10^6 eV.

meson. From the Greek *mésos*, meaning 'middle'. Mesons are a subclass of hadrons. They experience the strong nuclear force and are composed of quarks and antiquarks.

molecule. A fundamental unit of chemical substance formed from two or more atoms. A molecule of oxygen consists of two oxygen atoms, O_2. A molecule of water consists of two hydrogen atoms and one oxygen atom, H_2O.

MSSM. Acronym for the minimum supersymmetric standard model, the minimal extension of the conventional standard model of particle physics which accommodates supersymmetry, developed by in 1981 by Howard Georgi and Savas Dimopoulos.

M-theory. *See* string/M-theory.

muon. A second-generation lepton equivalent to the electron, with a charge -1, spin ½ (fermion), and mass 105.7 MeV/c^2. First discovered in 1936 by Carl Anderson and Seth Neddermeyer.

National Science Foundation (NSF). The NSF is an independent US federal government agency that supports education and research in science and engineering. It was founded in 1950 and in 2017 its annual budget was $7.5 billion. The NSF supports approximately 24 per cent of all federally sponsored basic research conducted by US colleges and universities.

neutrino. From Italian, meaning 'small neutral one'. Neutrinos are the chargeless, spin-½ (fermion) companions to the negatively charged electron, muon, and tau. The neutrinos are believed to possess very small masses, necessary to explain the phenomenon of neutrino oscillation, the quantum-mechanical mixing of neutrino flavours such that the flavour may change over time. Neutrino oscillation solves the solar neutrino problem—the problem that the numbers of neutrinos measured to pass through the Earth are inconsistent with the numbers of electron neutrinos expected from nuclear reactions occurring in the Sun's core. It was determined in 2001 that only 35 per cent of the neutrinos from the Sun are electron neutrinos—the balance are muon and tau neutrinos, indicating that the neutrino flavours oscillate as they travel from the Sun to the Earth.

neutron. An electrically neutral subatomic particle, first discovered in 1932 by James Chadwick. The neutron is a baryon consisting of one up and two down quarks with spin ½ and mass 939.6 MeV/c^2.

nucleus. The dense region at the core of an atom in which most of the atom's mass is concentrated. Atomic nuclei consist of varying numbers of protons and neutrons. The nucleus of a hydrogen atom consists of a single proton.

Pauli exclusion principle. Discovered by Wolfgang Pauli in 1925. The exclusion principle states that no two fermions may occupy the same quantum state (i.e. possess the same set of quantum numbers) simultaneously. For electrons in atoms, this means that only two electrons can occupy a single atomic orbital provided that they possess opposite spins.

perihelion. If a planet were to describe a circular orbit around the Sun, then there would obviously be no change in the distance between the Sun and the planet as it moves around its orbit. However, the planets of the Solar System describe elliptical orbits with the Sun at one focus. This means that the distance between the Sun and the planet does change. The perihelion is the point in the orbit at which the planet is closest to the Sun. The aphelion is the point at which the planet is furthest from the Sun. At its perihelion, the Earth is about 147.1 million kilometres from the Sun. At aphelion it is 152.1 million kilometres from the Sun.

photon. The elementary particle underlying all forms of electromagnetic radiation, including light. The photon is a massless, spin-1 boson which acts as the carrier of the electromagnetic force.

pion. A group of spin-0 mesons formed from up and down quarks and their antiquarks. These are π^+ (up-anti-down), π^- (down-anti-up), and π^0 (a mixture of up-anti-up and down-anti-down), with masses 139.6 MeV/c^2 ($\pi^\pm$) and 135.0 MeV/c^2 (π^0). The pions can be thought of as the 'carriers' of the strong force binding protons and neutrons inside the atomic nucleus, representing a kind of 'leakage' of the colour force binding quarks inside the protons and neutrons beyond the boundaries of these particles. First hypothesized by Japanese physicist Hideki Yukawa in 1935.

Planck constant. Denoted h. Discovered by Max Planck in 1900. The Planck constant is a fundamental physical constant which reflects the magnitudes of quanta in quantum theory. For example, the energies of photons are determined by their radiation frequencies according to the relation $E = h\nu$, energy equals Planck's constant multiplied by the radiation frequency. Planck's constant has the value 6.626×10^{-34} Joule-seconds.

Planck length. *See* Planck scale.

Planck scale. The Planck scale is defined by combining the values of the three fundamental physical constants central to quantum theory and relativity. These are the reduced Planck constant $\hbar$ (Planck's constant h divided by 2π—quantum theory), the speed of light c (special relativity), and Newton's gravitational constant, G (which makes an appearance in general relativity). The *Planck length*, given by $\sqrt{\hbar G / c^3}$, has the value 1.6×10^{-35} metres. The *Planck time*, $\sqrt{\hbar G / c^5}$, is the time taken for light to travel the Planck length, or 5.4×10^{-44} seconds. The *Planck mass*, $\sqrt{\hbar c / G}$, is about 2.2×10^{-8} kilograms, or about 0.02 milligrams. From the Planck length we can derive the Planck area and volume $\hbar G / c^3$ (2.6×10^{-70} square metres) and $\left(\hbar G / c^3\right)^{3/2}$ (4.2×10^{-105} cubic metres), respectively, and from the Planck mass we deduce the Planck energy using $E = mc^2$, giving $\sqrt{\hbar c^5 / G}$, which is about 2.0×10^9 joules. Combining the Planck mass and volume gives the Planck density, $c^5 / \hbar G^2$, which is about 5.2×10^{96} kilograms per cubic metre.

Planck time. *See* Planck scale.

positron. The antiparticle of the electron, denoted e^+, with a charge +1, spin ½ (fermion), and mass 0.511 MeV/c^2. The positron was the first antiparticle to be discovered, by Carl Anderson in 1932.

principle of general covariance. *See* general covariance.

principle of relativity. Requires that the equations relating to all the physical laws should have the same form irrespective of the choice of the frame of reference.

proton. A positively charged subatomic particle 'discovered' and so named by Ernest Rutherford in 1919. Rutherford actually identified that the nucleus of the hydrogen atom (which is a single proton) is a fundamental constituent of other atomic nuclei. The proton is a baryon consisting of two up and one down quarks with spin ½ and mass 938.3 MeV/c^2.

quantum. A fundamental, indivisible unit of properties such as energy and angular momentum. In quantum theory, such properties are recognized not to be continuously variable but to be organized in discrete packets or bundles, called quanta. In quantum field theory the use of the term is extended to include particles. Thus, the photon is the quantum particle of the electromagnetic field. This idea can be extended beyond the carriers of forces to include matter particles themselves. Thus, the electron is the quantum of the electron field, and so on. This is sometimes referred to as second quantization. In loop quantum gravity, space is also organized in discrete quanta of area and volume.

quantum chromodynamics (QCD). The quantum field theory of the colour force between quarks carried by a system of eight coloured gluons.

quantum electrodynamics (QED). The quantum field theory of the electromagnetic force between electrically charged particles, carried by photons.

quantum entanglement. A term coined by Erwin Schrödinger in 1935. Refers to a specific set of circumstances or physical processes in which the properties and behaviour of two or more quantum wave-particles are governed by a single wavefunction. Experiments on entangled particles (especially entangled photons) have been used to provide practical tests of local and crypto non-local hidden variable extensions of quantum theory based on Bell's and Leggett's inequalities.

quantum field. In classical field theory a 'force field' is ascribed a value at every point in spacetime and can be scalar (magnitude but no direction) or vector (magnitude and direction). The 'lines of force' made visible by sprinkling iron filings on a piece of paper held above a bar magnet provides a visual representation of such a field. In a quantum field theory, forces are conveyed by ripples in the field which form waves and— because waves can also be interpreted as particles—as quantum particles of the field. This idea can be extended beyond the carriers of forces (bosons) to include matter particles (fermions). Thus, the electron is the quantum of the electron field, and so on.

quantum geometrodynamics. The reformulation of general relativity as a set of classical Hamiltonian equations is sometimes referred to as geometrodynamics, by analogy with classical electrodynamics. When the classical field theory is then quantized, the result is quantum geometrodynamics, analogous to QED. *See also* canonical (approach to quantum gravity).

quantum number. The description of the physical state of a quantum system requires the specification of its properties in terms of total energy, linear and angular momentum, electric charge, etc. One consequence of the quantization of such properties is the appearance in this description of regular multiples of the associated quanta. The recurring integral or half-integral numbers which multiply the sizes of the quanta are called quantum numbers. When placed in a magnetic field, the electron spin may be oriented along or against the field lines of force, giving rise to 'spin-up' and 'spin-down' orientations characterized by the quantum numbers $+\frac{1}{2}$ and $-\frac{1}{2}$. Other examples include the principal number, n, which characterizes the energy levels of electrons in atoms, electric charge, quark colour charge, etc.

quantum probability. The wavefunctions of quantum wave-particles such as electrons are necessarily extended and delocalized through a region of space, for example in an orbital around the central proton in a hydrogen atom. The square of the amplitude of the wavefunction at some specific location is then related to the probability of 'finding' the electron at this point. The same principle applies to wavefunctions formed from superpositions. For example, if we form a wavefunction which is a mixture of spin-up and spin-down functions, then the probability that we will observe spin-up is given by the square of the amplitude of this component in the superposition. If spin-up is actually observed, the spin-down component 'disappears'. *See also* collapse of the wavefunction.

quark. The elementary constituents of hadrons. Hadrons are composed of triplets of spin-$\frac{1}{2}$ quarks (baryons) or combinations of quarks and anti-quarks (mesons). The quarks form three generations, each with different flavours. The up and down quarks, with electric charges $+\frac{2}{3}$ and $-\frac{1}{3}$ and masses of 1.8–3.0 and 4.5–5.3 MeV/c^2, respectively, form the first generation. Protons and neutrons are composed of up and down quarks. The second generation consists of the charm and strange quarks, with electric charges

+⅔ and −⅓ and masses of 1.28 GeV/c^2 and 95 MeV/c^2, respectively. The third generation consists of bottom and top quarks, with electric charges +⅔ and −⅓ and masses of 4.18 and 173 GeV/c^2, respectively. Quarks also carry colour charge, with each flavour of quark possessing red, green, or blue charges.

redshift. In the rainbow spectrum of colours, the energy of the light radiation increases from red through to violet. This means that red light has a lower frequency (and longer wavelength) than that of other colours. When the wavelength of radiation is increased as a result of the Doppler effect or the cosmological expansion of spacetime, the result is referred to as a 'redshift'. This doesn't mean that the radiation becomes 'redder', just that its wavelength is increased. For example, red light may be red-shifted into invisible infrared wavelengths. *See also* cosmological redshift.

renormalization. One consequence of introducing particles as the quanta of fields is that they may undergo self-interaction; i.e. they can interact with their own fields. This means that techniques, such as perturbation theory, used to solve the field equations tend to break down, as the self-interaction terms appear as infinite corrections. Renormalization was developed as a mathematical device used to eliminate these self-interaction terms, by redefining the parameters (such as mass and charge) of the field particles themselves. *See also* self-energy.

scale factor. *See* cosmic scale factor.

Schwarzschild solution/radius. German physicist Karl Schwarzschild was the first to provide exact solutions of Einstein's gravitational field equations, whilst serving in the German Army in 1916. The Schwarzschild solutions establish a fundamental boundary, called the *Schwarzschild radius*. A spherical mass, *m*, compressed to a radius less than the Schwarzschild radius (given by Gm/c^2, where c is the speed of light and G is the gravitational constant) will become a black hole—its escape velocity exceeds the speed of light.

self-energy. In a quantum field theory, particles are envisaged as fundamental fluctuations or vibrations of the field. One consequence is that particles may undergo self-interaction—they interact with their own fields. Such interactions increase the energy of the particle, by an amount called the self-energy. In early versions of the quantum field theory of electrons, the self-energy was found to be infinite. This problem was resolved by applying the techniques of renormalization.

singularity. The general theory of relativity allows for the possibility of infinite mass-energy density and gravitational field strength, such that spacetime curvature becomes infinite. The theory predicts such singularities at the centres of black holes and at the very 'beginning' of the Big Bang origin of the universe, though the geometries of the singularities in these instances are very different. These are *mathematical* singularities, which can be completely eliminated in theories such as loop quantum gravity in which space itself is quantized.

slow-roll inflation. The duration of the short burst of cosmic inflation and the way in which the early universe emerges from this depends very sensitively on the relationship between the density of energy that is stored in the inflaton field and the strength or magnitude of the field. As inflation is purely hypothetical, theorists are free to tweak its parameters so that the mechanism reproduces the large-scale structure of the universe that we see. In slow-roll inflation, the energy density declines only slightly as the field strength increases and the transition from false to true vacuum occurs slowly compared with the speed of expansion. This sustains the inflationary period long enough to solve the flatness and horizon problems without incurring additional problems. The energy density declines very sharply as the field strength approaches its value in the true vacuum, and inflation then ceases in an 'orderly' fashion. *See also* cosmic inflation, inflaton (field).

spacetime and spacetime metric. The distance between one position in a coordinate system and another can be determined from the values of the coordinates at these two positions. So, in a three-dimensional Euclidean space, if the positions are $l_1 = x_1 y_1 z_1$ and $l_2 = x_2 y_2 z_2$, the distance $\Delta l = l_2 - l_1$ can be found by applying Pythagoras' theorem: $\Delta l^2 = \Delta x^2 + \Delta y^2 + \Delta z^2$. This 'distance function' is often referred to as a *metric*. It has an important property: no matter how we define the coordinate system (no matter how we define x, y, and z), the metric will always be the same (mathematicians say that it is 'invariant'). We can extend Euclidean space to include a fourth dimension of time. To ensure that the resulting spacetime metric is invariant we need a structure such as $\Delta s^2 = \Delta (ct)^2 - \Delta x^2 - \Delta y^2 - \Delta z^2$, where s is a generalized spacetime interval, t is time, and c is the speed of light. We could swop these around and define Δs^2 such that the time interval is negative and the spatial intervals positive—so long as these are of opposite sign Δs^2 is invariant. The choice of signs is then simply a matter of convention.

special (theory of) relativity. Developed by Einstein in 1905, the special theory of relativity asserts that all motion is relative, and there is no unique or privileged frame of reference against which motion can be measured. All inertial frames of reference are equivalent—an observer stationary on Earth should obtain the same results from the same set of physical measurements as an observer moving with uniform velocity in a spaceship. Out go classical notions of absolute space, time, absolute rest, and simultaneity. In formulating the theory, Einstein assumed that the speed of light in a vacuum represents an ultimate speed which cannot be exceeded. The theory is 'special' only in the sense that it does not account for accelerated motion or gravity: this is covered in Einstein's general theory of relativity.

spectrum. Any physical property which has a range of possible values may be said to have a 'spectrum'. The most obvious example is the range of colours produced when light is passed through a prism or a collection of raindrops to produce a rainbow. The resulting spectrum may appear continuous (as in a rainbow) or it may be discrete, consisting of a set of specific values. The absorption or emission spectrum of hydrogen atoms exhibits a series of 'lines' corresponding to radiation frequencies that are absorbed or emitted by the atoms. The positions (frequencies) of these lines in the spectrum relate to the energies of the electron orbitals involved.

spin. All elementary particles exhibit a type of angular momentum called spin. Although the spin of the electron was initially interpreted in terms of 'self-rotation' (the electron spinning on its own axis, like a spinning top), spin is a relativistic phenomenon and has no counterpart in classical physics. Particles are characterized by their spin quantum numbers. Particles with half-integral spin quantum numbers are called fermions. Particles with integral spin quantum numbers are called bosons. Matter particles are fermions. Force particles are bosons.

spin connection. Transporting a vector quantity around a flat surface is straightforward, but becomes problematic when the surface is curved. Italian mathematician Tullio Levi-Civita developed a mathematical technique to keep track of the vector orientation which relies on the curvature of the surface. Instead of moving the vector we imagine that we're able to move the surface, and if the surface is spherical the technique is called the Levi-Civita connection on the sphere.

By the late 1970s, a similar logic had been applied to the parallel transport of vectors representing spin angular momentum, leading to systems of connections—called spin connections—that had found some useful applications in theories of solid state physics. Amitabha Sen realized that it might be possible to reformulate the ADM Hamiltonian form of general relativity, replacing the spacetime metric with spin connections from which the more familiar curved spacetime could then be derived. Sen's work formed the basis for Ashtekar's 'new variables'.

spinfoam. A spin network represents a static picture, so cannot be used to model a dynamics unfolding in time. To put time back into loop quantum gravity, we stack spin networks on top of each other, and trace the evolving relationships between the nodes and links. The two-dimensional graph of nodes and links now becomes a three-dimensional object in which the nodes become edges and the links become faces. The edges 'carry' a certain volume of space, and the faces 'carry' a certain area. The result is a spinfoam and Feynman's path-integral or sum-over-histories formulation of quantum mechanics is then used to trace the dynamics from start to finish. Spacetime then emerges as a superposition of spinfoams. The name 'spinfoam' acknowledges Wheeler's earlier intuition that spacetime at the Planck scale would be a 'quantum foam'.

spin network. Developed by Roger Penrose in the early 1970s as a way to represent elementary physical interactions without the need to assume a background spacetime. These interactions can be imagined to involve particles and fields, although this isn't actually all that important, and Penrose simply imagined them to involve 'objects' carrying spin angular momentum. A spin network consists of vertices or nodes, joined by lines which carry an integral number of quanta of spin angular momentum, given by $\frac{1}{2}\hbar$, where $\hbar$ is Planck's constant divided by 2π. Penrose showed that networks with sufficiently large total angular momentum serve to establish directions against which the relative orientation of other large networks can be measured, implying an emergent space. Loop quantum gravity independently arrives at very similar networks, in which the nodes are interpreted as the volumes of discrete quanta of space and the lines as discrete quanta of area where adjacent grains touch.

standard model, of Big Bang cosmology. *See* Λ-CDM model.

standard model, of particle physics. The currently accepted theoretical model describing matter particles and the forces between them, with the exception of gravity. The standard model consists of a collection of quantum field theories which describes three generations of quarks and leptons, the photon, W and Z particles, colour force gluons, and the Higgs boson.

strangeness. Identified as a characteristic property of particles such as the neutral lambda, neutral and charged sigma, and xi particles and the kaons. Strangeness was used together with electric charge and isospin to classify particles according to the 'Eightfold Way' by Murray Gell-Mann and Yuval Ne'eman. This property was subsequently traced to the presence in these composite particles of the strange quark.

strange quark. A second-generation quark with charge $-\frac{1}{3}$, spin $\frac{1}{2}$ (fermion), and a mass of 95 MeV/c^2. The property of 'strangeness' was identified as a characteristic of a series of relatively low-energy (low mass) particles discovered in the 1940s and 1950s by Murray Gell-Mann and independently by Kazuhiko Nishijima and Tadao Nakano. This property was subsequently traced by Gell-Mann and George Zweig to the presence in these composite particles of the strange quark.

string/M-theory. In string theory, the point-particles of the standard model of particle physics are replaced by extended one-dimensional strings. These were initially imagined to be 'lines of force' connecting quarks and antiquarks, but as the theory developed they became Planck-scale objects, with all the particles of the standard model presumed to be represented by the elementary vibrations of open and closed strings. Early versions of the theory were highly problematic, but considerable progress was made in the 'first superstring revolution' of 1984 by incorporating supersymmetry and assuming the strings exist in a complex space or manifold (called a Calabi–Yau space) with six 'hidden' spatial dimensions. It quickly emerged that there are several superstring theory variants, called Type I, Type IIA, Type IIB, and two different versions of so-called heterotic superstring theory. In the 'second superstring revolution' of 1995, all the variants were shown to be related through systems of 'dualities', and Edward Witten conjectured that these are all instances or limiting cases of a single, over-arching structure which he called M-theory. The theory is no longer limited to one-dimensional strings, but involves multi-dimensional branes.

strong force. The strong nuclear force, or colour force, binds quarks and gluons together inside hadrons and is described by quantum chromodynamics. The force that binds protons and neutrons together inside atomic nuclei (also referred to as the strong nuclear force) is thought to be the result of a 'leakage' of the colour force binding quarks inside the nucleons. *See also* colour force.

supergravity. The application of supersymmetry principles—involving the assumption of an essential spacetime symmetry between fermions and bosons—to general relativity produces a set of supergravity theories. The simplest examples involve the application of one supersymmetry and were first developed in 1976 by Dan Freedman, Sergio Ferrara, and Peter van Nieuwenhuizen and, independently, by Stanley Deser and Bruno Zumino. Assuming a supersymmetry removed some of the problems of renormalization (infinite contributions from radiative corrections involving the graviton could be partly offset by corresponding contributions from its supersymmetric partner, the gravitino), but it did not eliminate them entirely. Some excitement was generated in the early 1980s around an extended supergravity based on eight different kinds of supersymmetry, and although the renormalizability of this structure is still open to question today, the problems remain and the early interest in the theory has waned. *See also* supersymmetry (SUSY).

superinflation. In loop quantum cosmology, the universe cannot be compressed to a volume smaller than a single quantum of volume, and so a cosmic singularity is avoided. Instead the universe 'bounces'. As the universe emerges from the bounce the theory demands that it undergoes an unavoidable and very rapid burst of expansion, which continues as the mass-energy density falls from the critical density at the bounce, ρ_c, to half this value. This is 'superinflation', but it simply doesn't last long enough to explain the large-scale structure of the universe we see today. More conventional 'slow-roll' cosmic inflation must therefore be presumed. *See also* cosmic inflation, inflaton (field), slow-roll inflation.

superposition. In quantum mechanics, quantum entities can behave like particles and they can also behave like waves. But waves can be combined—they can be added together in a 'superposition'. Such combinations describe diffraction and interference effects. In a quantum measurement, it is necessary to form a superposition which contains contributions from the wavefunctions describing all the different possible outcomes. The square of the amplitude of each wavefunction in the superposition relates to the probability that the corresponding outcome will be observed. When the measurement is made, the wavefunction 'collapses' and all the other possible outcomes disappear.

supersymmetry (SUSY). A set of theoretical structures which assumes an essential spacetime symmetry between fermions and bosons. When applied to the standard model of particle physics, the asymmetry between matter particles (fermions) and force particles (bosons) is then explained in terms of a broken supersymmetry. At high energies (for example, the kinds of energies that prevailed during the very early stages of the Big Bang), supersymmetry would be unbroken and there would be perfect symmetry between fermions and bosons. Aside from the asymmetry between fermions and bosons, the broken supersymmetry predicts a collection of super-partners with a spin different by ½. In the simplest application of supersymmetry to the standard model, called the minimal supersymmetric standard model (MSSM), the partners of fermions are called sfermions. The partner of the electron is called the selectron; each quark is partnered by a corresponding squark. Likewise, for every boson there is a bosino. Supersymmetric partners of the photon, W and Z particles, and gluons are the photino, wino and zino, and gluinos, respectively. Supersymmetry resolves some of the problems with the current standard model, but unlike the symmetry between particles an antiparticles, it must further be assumed that all supersymmetric partners are massive and there is currently no satisfactory mechanism for supersymmetry-breaking that can readily explain this. Data collected from the Large Hadron Collider now effectively rule out the MSSM, and there is as yet no evidence for super-partners of any kind. *See also* MSSM, supergravity.

symmetry-breaking. Spontaneous symmetry-breaking occurs whenever the lowest-energy state of a physical system has lower symmetry than higher-energy states. As the system loses energy and settles to its lowest-energy state, the symmetry spontaneously reduces, or 'breaks'. For example, a pencil perfectly balanced on its tip is symmetrical, but under the influence of the background environment (such as small currents or air) it will topple over to give a more stable, lower energy, less symmetrical state with the pencil lying along one specific direction.

tera. A prefix denoting trillion. A tera electron volts (TeV) is a trillion electron volts, 10^{12} eV, or 1,000 GeV.

top quark. Also sometimes referred to as the 'truth' quark. A third-generation quark with charge $+\frac{2}{3}$, spin $\frac{1}{2}$ (fermion), and a mass of 173 GeV/c^2. It was discovered at Fermilab in 1995.

trillion. A thousand billion or a million million, 10^{12} or 1,000,000,000,000.

uncertainty principle. Discovered by Werner Heisenberg in 1927. The uncertainty principle states that there is a fundamental limit to the precision with which it is possible to measure pairs of 'conjugate' observables, such as position and momentum and energy and the rate of change of energy with time. The principle can be traced to the fundamental duality of wave and particle behaviour in quantum objects.

vacuum energy. *See* cosmological constant.

vacuum expectation value. In quantum theory, the magnitudes of observable quantities such as energy are given as the so-called expectation (or average) values of the quantum mechanical operators which correspond to the observables. The operators are mathematical functions which operate on, and thereby change, the wavefunctions. The vacuum expectation value is the expectation value of the operator in a vacuum. The Higgs field has a non-zero vacuum expectation value, and this breaks the symmetry of the electroweak force to produce distinct electromagnetic and weak nuclear forces.

virtual particle. Virtual particles were introduced in early quantum field theories as a way of representing interactions between 'real' particles involving forces. Although by definition virtual particles are never 'seen', their physical effects have been measured (and you can feel the effects of the exchange of virtual photons whenever you try to push the south poles of two bar magnets together). Virtual particles are transient fluctuations in a quantum field whose lifetimes are limited by the uncertainty principle. Although they obey the laws of physics (such as energy or momentum conservation), they may have distinctly different properties compared with their 'real' counterparts. The longer a virtual particle persists, the closer it will resemble its real counterpart.

W, Z particles. Elementary particles which carry the weak nuclear force. The W particles are spin-1 bosons with unit positive and negative electrical charge (W^+, W^-) and masses of 80.4 GeV/c^2. The Z^0 is an electrically neutral spin-1 boson with mass 91.2 GeV/c^2. The W and Z particles gain mass through the Higgs mechanism.

wavefunction. The mathematical description of matter particles such as electrons as 'matter waves' leads to equations characteristic of wave motion. Such wave equations feature a wavefunction whose amplitude and phase evolves in space and time. The wavefunctions of the electron in a hydrogen atom form characteristic three-dimensional patterns around the nucleus called orbitals. Wave mechanics—an expression of quantum mechanics in terms of matter waves—was first elucidated by Erwin Schrödinger in 1926.

wavefunction collapse. *See* collapse of the wavefunction.

wave–particle duality. A fundamental property of all quantum particles, which exhibit both delocalized wave behaviour (such as diffraction and interference) and localized particle behaviour depending on the type of apparatus used to make measurements on them. First suggested as a property of matter particles such as electrons by Louis de Broglie in 1923.

weak force/interaction. The weak force is so called because it is considerably weaker than both the strong and electromagnetic forces, in strength and range. The weak force affects both quarks and leptons and weak force interactions can change quark and lepton flavour: for example, turning an up quark into a down quark and an electron into an electron neutrino. The weak force was originally identified as a fundamental force from studies of beta radioactive decay. Carriers of the weak force are the W and Z particles. The weak force was combined with electromagnetism in the quantum field theory of the electroweak force by Steven Weinberg and Abdus Salam in 1967–8.

Wilson loops. Introduced in 1974 by American theorist Kenneth Wilson and independently by Russian theorist Alexander Polyakov in an attempt to formulate a version of quantum chromodynamics that could be solved analytically. The reformulation involves focusing on the lines of colour force: the action of a quantum-mechanical operator then results in an elementary excitation of the quantum field localized in a loop. In this way, the lines or loops of force become primary, and the quantum field is a secondary derivative of these.

Yang–Mills field theory. A form of quantum field theory developed in 1954 by Chen Ning Yang and Robert Mills. Yang–Mills field theories underpin all the components of the current standard model of particle physics.

ENDNOTES

When referencing the primary scientific literature I've tried as much as possible to provide the original published article *and* the relevant preprint, posted on the online preprint archive managed by Cornell University. The preprints can be accessed free of charge from the arXiv home page—http://arxiv.org—and typing the article identifier in the search window. Where they are given, direct quotes are typically derived from the preprint.

PREFACE

1. Joseph Conlon's readable and engaging book *Why String Theory?*, published in 2016, offers a similar assessment of the evidence for string theory. Chapter 7, titled 'Direct Experimental Evidence for String Theory', consists of just a single sentence, which reads 'There is no direct experimental evidence for string theory.' Joseph Conlon, *Why String Theory?*, CRC Press, Boca Raton, FL, 2016, p. 107.

2. This is my view and, of course, not all scientists agree. For example, South African cosmologist George Ellis argues that the greatest scientific challenge is consciousness (personal communication, 25 October 2017).

3. Jakub Mielczarek and Tomasz Trzesniewski, 'Towards the Map of Quantum Gravity', arXiv:hep-th/1708.07445v1, 24 August 2017.

4. Marcus Chown, *The Ascent of Gravity: The Quest to Understand the Force That Explains Everything*, Weidenfeld & Nicolson, London, 2017. The footnote in question appears on p. 252.

5. Carlo Rovelli, 'Loop Quantum Gravity: The First 25 Years', *Classical and Quantum Gravity*, **28** (2011) 153002; arXiv:gr-qc/1012.4707v5, 28 January 2012, p. 20.

6. Those with a background in physics might want to consult a recent volume of reviews written mainly by young LQG researchers: Abhay Ashtekar and Jorge Pullin (eds), *Loop Quantum Gravity: The First 30 Years*, World Scientific, Singapore, 2017.

PROLOGUE

1. Lee Smolin, *The Life of the Cosmos*, Oxford University Press, Oxford, 1997, pp. 7–8.
2. Albert Einstein, in Paul Arthur Schilpp (ed.), *Albert Einstein: Philosopher-Scientist*, Harper & Row, New York, 1959, p. 3.
3. Ibid., p. 5.
4. Albert Einstein, letter to F. Lentz, 20 August 1949, quoted in Alice Calaprice (ed.), *The Ultimate Quotable Einstein*, Princeton University Press, Princeton, NJ, 2011, p. 19.
5. Smolin, *The Life of the Cosmos*, pp. 7–8.
6. On *Desert Island Discs*, first broadcast on 2 July 2017 on BBC Radio 4, Rovelli said: 'I grew up in a very lovely family, with a very loving Italian mother. I was an only child, completely immersed in this maternal love, which was great—it gave me security and it gave me strength, but it was also a prison from which I had to escape at some point.'
7. Carlo Rovelli, personal communication, 19 August 2017.
8. Carlo Rovelli, *What is Time? What is Space?*, manuscript translated by J. C. van den Berg, p. 2. Published by de Renzo Editore, Rome, 2006.
9. Albert Einstein, *Relativity: The Special and the General Theory*, 100th anniversary edition, Princeton University Press, Princeton, NJ, 2015, p. 156.
10. Rovelli, *What is Time?*, p. 3.

CHAPTER 1: THE LAWS OF PHYSICS ARE THE SAME FOR EVERYONE

1. Albert Einstein, *Relativity: The Special and the General Theory*, 100th anniversary edition, Princeton University Press, Princeton, NJ, 2015, p. 155.
2. J. R. R. Tolkein, *The Lord of the Rings: The Two Towers*, Harper Collins, London, 1997, p. 586.
3. See 'Einstein', *Metromnia*, National Physical Laboratory, Issue 18, Winter 2005.
4. Carlo Rovelli, *Reality is Not What it Seems: The Journey to Quantum Gravity*, Allen Lane, London, 2016, pp. 59–60.

5. Albert Einstein, 'Does the Inertia of a Body Depend on its Energy Content?', *Annalen der Physik*, **18** (1905), 639–41. This paper is translated and reproduced in John Stachel (ed.), *Einstein's Miraculous Year: Five Papers that Changed the Face of Physics*, centenary edition, Princeton University Press, Princeton, NJ, 2005. The quote appears on p. 164.

6. Isaac Newton, *Mathematical Principles of Natural Philosophy*, first American edition, translated by Andrew Motte, published by Daniel Adee, New York, 1845, p. 506.

CHAPTER 2: THERE'S NO SUCH THING AS THE FORCE OF GRAVITY

1. Albert Einstein, in Paul Arthur Schilpp (ed.), *Albert Einstein: Philosopher-Scientist*, Harper & Row, New York, 1959, p. 27.

2. Albert Einstein, *Relativity: The Special and the General Theory*, 100th anniversary edition, Princeton University Press, Princeton, NJ, 2015, pp. 163–4.

3. Isaac Newton, *Mathematical Principles of Natural Philosophy*, first American edition, translated by Andrew Motte, published by Daniel Adee, New York, 1845, p. 81.

4. Einstein was forever indebted to Mach for his approach to physics, but not his aggressively empiricist approach to philosophy, which among other things led him reject the existence of atoms. Einstein once commented that 'Mach was as good at mechanics as he was wretched at philosophy'. Albert Einstein, quoted in Abraham Pais, *Subtle is the Lord: The Science and the Life of Albert Einstein*, Oxford University Press, Oxford, 1982. The quote appears on p. 283.

5. Albert Einstein, in the 'Morgan manuscript', quoted ibid., p. 178.

6. Albert Einstein, 'How I Created the Theory of Relativity', lecture delivered at Kyoto University, 14 December 1922, translated by Yoshimasa A. Ono, *Physics Today*, August 1982, p. 47.

7. Lee Smolin, personal communication, 7 September 2017.

8. John Wheeler, with Kenneth Ford, *Geons, Black Holes and Quantum Foam: A Life in Physics*, W. W. Norton, New York, 1998, p. 228.

9. Ibid., p. 235.

10. Albert Einstein, letter to Paul Ehrenfest, 17 January 1916, quoted in Robert E. Kennedy, *A Student's Guide to Einstein's Major Papers*, Oxford University Press, Oxford, 2012. The quote appears on p. 214.

11. 'Einstein', *Metromnia*, National Physical Laboratory, Issue 18, Winter 2005, p. 3.

12. Neil Ashby, 'Relativity and the Global Positioning System', *Physics Today*, May 2002, p. 42.

13. Lee Smolin, personal communication, 7 September 2017.

14. Carlo Rovelli, *Reality is Not What it Seems: The Journey to Quantum Gravity*, Allen Lane, London, 2016, p. 71.

15. Albert Einstein, in A. J. Knox, Martin J. Klein, and Robert Schulmann (eds), *The Collected Papers of Albert Einstein*, vol. 6, *The Berlin Years: Writings 1914–1917*, Princeton University Press, Princeton, NJ, 1996, p. 153.

16. Rovelli, *Reality is Not What it Seems*, p. 73.

CHAPTER 3: WHY NOBODY UNDERSTANDS QUANTUM MECHANICS

1. Werner Heisenberg, *Physics and Beyond: Memories of a Life in Science*, George Allen & Unwin, London, 1971, p. 73.

2. Einstein wrote: 'The theory produces a good deal but hardly brings us closer to the secret of the Old One. I am at all events convinced that *He* does not play dice'. Albert Einstein, letter to Max Born, 4 December 1926. Quoted in Abraham Pais, *Subtle is the Lord: The Science and the Life of Albert Einstein*, Oxford University Press, Oxford 1982, p. 443.

3. P. A. M. Dirac, *The Principles of Quantum Mechanics*, 4th edition, Oxford University Press, Oxford, 1958, pp. viii–ix.

4. Paul Dirac, interview 6 May 1963, *Archives for the History of Quantum Physics*, p. 6, quoted by Graham Farmelo, *The Strangest Man: The Hidden Life of Paul Dirac, Quantum Genius*, Faber & Faber, London, 2009, p. 44.

5. Paul Dirac, in G. Holton and Y. Elkana (eds), *Albert Einstein: Historical and Cultural Perspectives*, Princeton University Press, Princeton, NJ, 1982, p. 84. Quoted by Farmelo, *The Strangest Man*, p. 137.

6. Carlo Rovelli, personal communication, 29 January 2017.

7. Richard Feynman, *The Character of Physical Law*, MIT Press, Cambridge, MA, 1967, p. 129. The italics are mine.

8. Richard P. Feynman, Robert B. Leighton, and Matthew Sands, *The Feynman Lectures on Physics*, volume III, Addison-Wesley, Reading, MA, 1965, p. 1.

9. Heisenberg wrote: 'I remember discussions with Bohr which went through many hours till very late at night and ended almost in despair; and when at the end of the discussion I went alone for a walk in the neighbouring park I repeated to myself again and again the question: Can nature possibly be as absurd as it seemed...?' Werner Heisenberg, *Physics and Philosophy: The Revolution in Modern Science*, Penguin, London, 1989 (first published 1958), p. 30.

10. Albert Einstein, letter to Max Born, 1952. Quoted in John S. Bell, *Proceedings of the Symposium on Frontier Problems in High Energy Physics, Pisa, June 1976*, pp. 33–45. This paper is reproduced in J. S. Bell, *Speakable and Unspeakable in Quantum Mechanics*, Cambridge University Press, Cambridge, 1987, pp. 81–92. The quote appears on p. 91.

11. Albert Einstein, Boris Podolsky, and Nathan Rosen, 'Can Quantum-Mechanical Description of Physical Reality be Considered Complete?', *Physical Review*, **47**, 1935, 777–80. This paper is reproduced in John Archibald Wheeler and Wojciech Hubert Zurek (eds), *Quantum Theory and Measurement*, Princeton University Press, Princeton, 1983, p. 141.

12. Bell wrote: 'If the [hidden variable] extension is local it will not agree with quantum mechanics, and if it agrees with quantum mechanics it will not be local. This is what the theorem says.' John Bell, 'Locality in Quantum Mechanics: Reply to Critics', *Epistemological Letters*, November 1975, 2–6. This paper is reproduced in Bell, *Speakable and Unspeakable in Quantum Mechanics*, pp. 63–6. This quote appears on p. 65.

13. Bell's theorem can be expressed in the form of an *inequality* between sets of experimental results. For example, for one specific experimental arrangement, the generalized form of Bell's inequality demands a value that cannot be greater than 2. Quantum theory predicts a value of $2\sqrt{2}$, or 2.828. Aspect and his colleagues obtained the result 2.697 ± 0.015. In other words, the experimental result exceeded the maximum limit predicted by Bell's inequality by almost fifty times the experimental error, a powerful, statistically significant violation.

14. Lee Smolin, personal communication, 7 September 2017.

15. A. J. Leggett, 'Non-local Hidden Variable Theories and Quantum Mechanics: An Incompatibility Theorem', *Foundations of Physics*, **33** (2003), 1474–5.

16. Like Bell's theorem, Leggett's crypto non-local hidden variables theory can also be tested through an inequality between experimental results. For a specific experimental arrangement, the whole class of crypto non-local hidden variable theories predicts a maximum value for the Leggett inequality of 3.779. Quantum theory violates this inequality, predicting a value of 3.879, a difference of less than 3 per cent. The experimental result was 3.852 ± 0.023, a violation of the Leggett inequality by more than three times the experimental error. See Simon Gröblacher, Tomasz Paterek, Rainer Kaltenbaek, Caslav Brukner, Marek Zukowski, Markus Aspelmeyer, and Anton Zeilinger, 'An Experimental Test of Non-local Realism', *Nature*, **446** (2007), 875; arXiv:quant-ph/0704.2529v2, 6 August 2007.

17. In these experiments for a specific arrangement the maximum value allowed by the Leggett inequality is 1.78868, compared with the quantum theory prediction of 1.93185. The experimental result was 1.9323 ± 0.0239, a violation of the inequality by more than six times the experimental error. See J. Romero, J. Leach, B. Jack, S. M. Barnett, M. J. Padgett, and S. Franke-Arnold, 'Violation of Leggett Inequalities in Orbital Angular Momentum Subspaces', *New Journal of Physics*, **12** (2010), 123007.

18. For a summary, see Alain Aspect, 'Closing the Door on Einstein and Bohr's Quantum Debate', *Physics*, **8**, 123, 16 December 2015, available at http://physics.aps.org/articles/v8/123.

CHAPTER 4: MASS AIN'T WHAT IT USED TO BE

1. Paul A. M. Dirac, *Lectures on Quantum Mechanics*, Dover, New York, 2001 (first published 1964), p. 1.

2. It might help to think about electron spin like this. Make a Möbius band by taking a length of tape, twisting it once and joining the ends together so the band is continuous and seamless. What you have is a ring of tape with only one 'side' (it doesn't have distinct outside and inside surfaces). Now picture yourself walking along this band. You'll find that, to get back to where you start, you need to walk twice around the ring.

3. Richard P. Feynman, *QED: The Strange Story of Light and Matter*, Penguin Books, New York, 1990 (first published 1985), p. 7.

4. We can get a rough handle on the relationship between the range of a force and the mass of force-carrying particles by combining Heisenberg's uncertainty principle and Einstein's equation $E = mc^2$. A bit of algebraic manipulation then tells us that the range of the force is *inversely proportional* to the mass of the force-carrier (so the range of a force carried by massless particles like photons is essentially infinite). Yukawa assumed that the range of the strong nuclear force must be on the order of the radius of a proton, from which he deduced the mass of the force carrier to be about 200 times the mass of the electron.

5. In his book *The God Particle*, Lederman gave two reasons for this name: 'One, the publisher wouldn't let us call it the Goddamn Particle, though that might be a more appropriate title, given its villainous nature and the expense it is causing. And two, there is a connection, of sorts, to another book, a much older one'. Leon Lederman (with Dick Teresi), *The God Particle: If the Universe is the Answer, What is the Question?*, Bantam Press, London, 1993, p. 22.

6. Carlo Rovelli, *Reality is Not What it Seems: The Journey to Quantum Gravity*, Allen Lane, London, 2016, p. 110.

7. In his Foreword to my book *Higgs*, Weinberg wrote: 'Rather, I did not include quarks in the theory simply because in 1967 I just did not believe in quarks. No-one had ever observed a quark, and it was hard to believe that this was because quarks are much heavier than observed particles like protons and neutrons, when these observed particles were supposed to be made of quarks.' See Jim Baggott, *Higgs: The Invention and Discovery of the 'God' Particle*, Oxford University Press, Oxford, 2012, p. xx.

8. Gerardus 't Hooft, in *Les Prix Nobel: The Nobel Prizes 1999*, ed. Tore Frängsmyr, Nobel Foundation, Stockholm, 2000. Available online at https://www.nobelprize.org/nobel_prizes/physics/laureates/1999/thooft-bio.html/

CHAPTER 5: HOW TO FUDGE THE EQUATIONS OF THE UNIVERSE

1. Albert Einstein, letter to Paul Ehrenfest, 4 February 1917, quoted in Abraham Pais, *Subtle is the Lord: The Science and the Life of Albert Einstein*, Oxford University Press, Oxford, 1982. The quote appears on p. 285.

2. Mark A. Peterson, 'Dante and the 3-Sphere', *American Journal of Physics*, **47** (1979), 1033. Dante's solution was to extend the model to a fourth dimension determined by the *speed of revolution* of his heavenly spheres.

3. Newton wrote: 'and lest the systems of the fixed stars should, by their gravity, fall on each other mutually, he [God] hath placed those systems at immense distances one from another.' Isaac Newton, *Mathematical Principles of Natural Philosophy*, first American edition, trans. Andrew Motte, published by Daniel Adee, New York, 1845, p. 504.

4. Albert Einstein, 'Cosmological Considerations in the General Theory of Relativity', *Proceedings of the Prussian Academy of Sciences*, **142** (1917). Quoted in Walter Isaacson, *Einstein: His Life and Universe*, Simon & Shuster, New York, 2007, p. 255.

5. Hubble's law is $v = H_0 D$, where v is the speed of the galaxy, H_0 is Hubble's constant as measured in the present time, and D is the so-called 'proper distance' of the galaxy measured from the Earth, such that the speed is then given simply as the rate of change of this distance. Although it is often referred to as a 'constant', in truth the Hubble parameter varies with time depending on assumptions regarding the rate of expansion of the universe.

6. George Gamow wrote: 'When I was discussing cosmological problems with Einstein he remarked that the introduction of the cosmological term was the biggest blunder he ever made in his life.' George Gamow, *My World Line: An Informal Autobiography*, Viking Press, New York, 1970, p. 149. Quoted in Walter Isaacson, *Einstein: His Life and Universe*, Simon & Shuster, New York, 2007, pp. 355–6.

7. Alan H. Guth, *The Inflationary Universe: The Quest for a New Theory of Cosmic Origins*, Vintage, London, 1998, p. 86.

8. S. D. M. White and M. J. Rees, 'Core Condensation in Heavy Halos: A Two-Stage Theory for Galaxy Formation and Clustering', *Monthly Notices of the Royal Astronomical Society*, **183** (1978), 341–58.

9. Lemaître wrote: 'Everything happens as though the energy in vacuo would be different from zero.' G. Lemaître, 'L'univers en Expansion', *Annales de la Société Scientifique de Bruxelles, Serie A*, **53** (1933), 51–85. Quoted in Harry Nussbaumer and Lydia Bieri, *Discovering the*

Expanding Universe, Cambridge University Press, Cambridge, 2009, p. 171.

10. Carlo Rovelli, *Reality is Not What it Seems: The Journey to Quantum Gravity*, Allen Lane, London, 2016, pp. 178–9.

CHAPTER 6: TO GET THERE I WOULDN'T START FROM HERE

1. Albert Einstein, 'Approximative Integration of the Field Equations of Gravitation', *Preussische Akademie der Wissenschaften (Berlin) Sitzungsberichte* (1916), 688–96. Quoted in Gennady E. Gorelik and Viktor Ya. Frenkel, *Matvei Petrovich Bronstein and Soviet Theoretical Physics in the Thirties*, Birkhäuser Verlag, Basel, 1994. The quote appears on p. 86.

2. The Planck scale is defined by combining the values of the three fundamental physical constants that are central to quantum theory and relativity. These are the reduced Planck constant $\hbar$ (Planck's constant h divided by 2π—quantum theory), the speed of light c (special relativity), and Newton's gravitational constant, G (which makes an appearance in general relativity). The *Planck length*, given by $\sqrt{\hbar G/c^3}$, has the value 1.6×10^{-35} metres. The *Planck time*, $\sqrt{\hbar G/c^5}$, is the time taken for light to travel the Planck length, or 5.4×10^{-44} seconds. The *Planck mass*, $\sqrt{\hbar c/G}$, is about 2.2×10^{-8} kilograms, or about 0.02 milligrams. From the Planck mass we deduce the Planck energy using $E = mc^2$, giving $\sqrt{\hbar c^5/G}$, which is about 2.0×10^9 joules, or about 0.5 mega-watt hours. This is approximately equivalent to the amount of electricity consumed by an average American in 12 days.

3. Paul Dirac, *Süddeutsche Zeitung*, 18 October 1963, quoted in Julian Barbour, *The End of Time: The Next Revolution in Our Understanding of the Universe*, Phoenix, London, 2000. The quote appears on p. 2.

4. Paul A. M. Dirac, *Lectures on Quantum Mechanics*, Dover, New York, 2001 (first published 1964), p. 4.

5. Bryce DeWitt, 'Quantum Theory of Gravity. I. The Canonical Theory', *Physical Review*, **160** (1967), 1119.

6. Ibid., 1131.

7. Carlo Rovelli, *What is Time? What is Space?*, manuscript translated by J. C. van den Berg, p. 11. Published by de Renzo Editore, Rome, 2006.

8. Richard Feynman, Interviews and conversations with Jagdish Mehra, in Pasadena, California, January 1988, quoted in Jagdish Mehra, *The Beat of a Different Drum: The Life and Science of Richard Feynman*, Oxford University Press, Oxford, 1994. The quote appears on p. 507.

9. Steven Weinberg, *Gravitation and Cosmology*, John Wiley & Sons, New York, 1972. Quoted in Abhay Ashtekar, 'The Winding Road to Quantum Gravity', *Current Science*, **89** (2005), 2066.

10. Lee Smolin, *Three Roads to Quantum Gravity: A New Understanding of Space, Time and the Universe*, Phoenix, London, 2001, p. 6.

11. Rovelli, *What is Time?*, p. 6.

CHAPTER 7: A GIFT FROM THE DEVIL'S GRANDMOTHER

1. Kenneth R. Miller, quoted by Carl Zimmer, *New York Times*, 8 April 2016, available at https://www.nytimes.com/2016/04/09/science/in-science-its-never-just-a-theory.html?_r=2

2. Howard Georgi, interview with Robert Crease and Charles Mann, 29 January 1985. Quoted in Robert P. Crease and Charles C. Mann, *The Second Creation: Makers of the Revolution in Twentieth-Century Physics*, Rutgers University Press, New Brunswick, NJ, 1986, p. 400.

3. Sheldon L. Glashow, with Ben Bova, *Interactions: A Journey through the Mind of a Particle Physicist and the Matter of This World*, Warner Books, New York, 1988, p. 309.

4. Stephen P. Martin, 'A Supersymmetry Primer', version 6, arXiv:hep-ph/9709356, September 2011, p. 5.

5. S. W. Hawking, 'Is the End in Sight for Theoretical Physics?', *Physics Bulletin*, **32** (1981), 17. This is an abbreviated version of Hawking's inaugural lecture, delivered on 29 April 1980.

6. Albert Einstein, letter to Erwin Schrödinger, 20 May 1946, quoted in Walter Moore, *Schrödinger: Life and Thought*, Cambridge University Press, Cambridge, UK, 1989, p. 426.

7. Amitabha Sen, 'Gravity as a Spin System', *Physics Letters B*, **119** (1982), 91.

8. Abhay Ashtekar, *Ashtekar Variables*, available at http://www.scholarpedia.org/article/Ashtekar_variables.

9. Lee Smolin, *Three Roads to Quantum Gravity: A New Understanding of Space, Time and the Universe*, Phoenix, London, 2001, p. 122.

10. Ibid., pp. 125–6.

CHAPTER 8: OUR SECOND OR THIRD GUESS SOLVED THE EQUATIONS EXACTLY

1. Leonard Susskind, *The Landscape: A Talk with Leonard Susskind, Edge. org*, April 2003. Available at https://www.edge.org/conversation/leonard_susskind-the-landscape.

2. John Schwarz, interview with Sara Lippincott, 21 and 26 July, 2000, Oral History Project, California Institute of Technology Archives, 2002, p. 17.

3. Lattice QCD and combined QED calculations of the small mass difference between the proton and the neutron were reported in March 2015: Sz. Borsanyi et al., 'Ab initio Calculation of the Neutron-Proton Mass Difference', *Science*, **347** (2015), 1452–5, arXiv:hep-lat/1406.4088v2, 7 April 2015. See also the commentary by Frank Wilczek, 'A Weighty Mass Difference', *Nature*, **520** (2015), 303–4, and Jim Baggott, *Mass: The Quest to Understand Matter from Greek Atoms to Quantum Fields*, Oxford University Press, Oxford, 2017.

4. Lee Smolin, *Three Roads to Quantum Gravity: A New Understanding of Space, Time and the Universe*, Phoenix, London, 2001, p. 119.

5. Ibid., p. 40.

6. Carlo Rovelli, *What is Time? What is Space?*, manuscript translated by J. C. van den Berg, p. 13. Published by de Renzo Editore, Rome, 2006.

7. Brian Greene, *The Hidden Reality: Parallel Universes and the Deep Laws of the Cosmos*, Allen Lane, London, 2011, p. 91.

8. Strominger had shown that the 'compactification' of the six extra spatial dimensions could be achieved with vastly more spaces than those classified as Calabi–Yau spaces: A. Strominger, 'Superstrings with Torsion', *Nuclear Physics B*, **274** (1986), 253–84.

9. Richard Feynman, in P. C. W. Davies and Julian Brown, eds, *Superstrings: A Theory of Everything?*, Cambridge University Press, Cambridge, UK, 1988, pp. 194–5.

10. Sheldon L. Glashow, with Ben Bova, *Interactions: A Journey through the Mind of a Particle Physicist and the Matter of This World*, Warner Books, New York, 1988, p. 334.

CHAPTER 9: I USED EVERY AVAILABLE KEY
RING IN VERONA

1. Ted Jacobson and Lee Smolin, 'Nonperturbative Quantum Geometry', *Nuclear Physics B*, **299** (1988), 295. This quote appears on p. 343.

2. Ted Jacobson, personal communication, 8 December 2017.

3. Carlo Rovelli, quoted by Manjit Kumar, 'Theory of (Almost) Everything', *Literary Review*, December 2016, available at https:// literaryreview.co.uk/theory-of-almost-everything.

4. Lee Smolin, *Three Roads to Quantum Gravity: A New Understanding of Space, Time and the Universe*, Phoenix, London, 2001, p. 129.

5. Carlo Rovelli, *What is Time? What is Space?*, manuscript translated by J. C. van den Berg, p. 13. Published by de Renzo Editore, Rome, 2006.

6. Smolin, *Three Roads to Quantum Gravity*, p. 129.

7. Ibid., p. 130.

8. Abhay Ashtekar, personal communication, 11 December 2017.

9. Carlo Rovelli, personal communication, 15 December 2017.

10. Smolin, *Three Roads to Quantum Gravity*, p. 131.

11. Carlo Rovelli and Lee Smolin, 'Knot Theory and Quantum Gravity', *Physical Review Letters*, **61** (1988), 1158.

12. Marcia Bartusiak, 'Loops of Space', *Discover*, April 1993, p. 66.

13. Lee Smolin, quoted in ibid., p. 66.

14. Rovelli, *What is Time? What is Space?*, p. 16.

15. Carlo Rovelli, quoted in Bartusiak, 'Loops of Space', p. 66.

16. Lee Smolin, 'Recent Developments in Nonperturbative Quantum Gravity', in *Quantum Gravity and Cosmology, Proceedings of the 1991 GIFT International Seminar on Theoretical Physics*, held in Saint Feliu de Guixols, Catalonia, Spain, World Scientific, Singapore, 1992. See also arXiv:hep-th/9202022v1, 7 February 1992.

17. Lee Smolin, personal communication, 29 December 2017.

18. Abhay Ashtekar, Carlo Rovelli, and Lee Smolin, 'Weaving a Classical Geometry with Quantum Threads', *Physical Review Letters*, **69** (1992), 237. See also arXiv:hep-th/9203079v1, 30 March 1992. Ashtekar actually took weaving lessons so that he could better understand the process.

19. Carlo Rovelli, personal communication, 26 December 2017.

20. The spectrum of area is proportional to $8\pi l_p^2 \sqrt{j(j+1)}$, where l_p is the Planck length, 1.6×10^{-35} metres. In fact, Rovelli had been

puzzling over a formula which featured the square root $\sqrt{p^2 + 2p}$, where p is an integer. Junichi Iwasaki, then a student, stormed into his office and demanded that he rewrite the equation setting $j = \frac{1}{2}p$. This transformed the formula into an expression immediately recognizable from the quantum mechanics of angular momentum.

21. Roger Penrose, *The Road to Reality: A Complete Guide to the Laws of the Universe*, Vintage, London, 2005, p. 947. The italics are Penrose's.

22. Roger Penrose, *The Emperor's New Mind: Concerning Computers, Minds and the Laws of Physics*, Vintage, London, 1990, p. 341.

23. Roger Penrose, 'Angular Momentum: An Approach to Combinatorial Space-time', first published in Ted Bastin (ed.), *Quantum Theory and Beyond*, Cambridge University Press Cambridge, UK, 1971, pp. 151–80. This book is now out of print but you can retrieve a copy of this chapter from John Baez's website: http://math.ucr.edu/home/baez/penrose/. This quote appears on p. 2.

24. In fact, Penrose's integers n are equal to $2j$, where j is the total spin angular momentum quantum number. The total spin angular momentum along a line in the spin network is then given by $\hbar\sqrt{j(j+1)}$, exactly analogous with the area spectrum of LQG. Of course, Penrose's n is the same as Rovelli's original p (see endnote 20).

25. Penrose, 'Angular Momentum', p. 4.

26. Lee Smolin, personal communication, 29 December 2017.

27. The two papers are: Carlo Rovelli and Lee Smolin, 'Discreteness of Area and Volume in Quantum Gravity', *Nuclear Physics B*, **442** (1995), 593–619 (arXiv:gr-qc/9411005v1, 2 November 1994); and Carlo Rovelli and Lee Smolin, 'Spin Networks and Quantum Gravity', *Physical Review D*, **52** (1995), 5743 (arXiv:gr-qc/9505006, 4 May 1995).

28. Carlo Rovelli, personal communication, 15 December 2017.

CHAPTER 10: IS THERE REALLY NO TIME LIKE THE PRESENT?

1. Carlo Rovelli and Lee Smolin, 'Discreteness of Area and Volume in Quantum Gravity', *Nuclear Physics B*, **442** (1995), 593–619; arXiv: gr-qc/9411005v1, 2 November 1994. The quote appears on p. 5 of the archived paper.

2. See, for example, Helge Kragh, *Higher Speculations: Grand Theories and Failed Revolutions in Physics and Cosmology*, Oxford University Press, Oxford, 2011, pp. 300–1.

3. Steven Weinberg expressed this sentiment to me as follows: 'String theory still looks promising enough to be worth further effort. I wouldn't say this if there were a more promising alternative available, but there isn't. We are in the position of a gambler who is warned not to get into a poker game because it appears to be crooked; he explains that he has no choice, because it is the only game in town.' Personal communication, 13 January 2013.

4. Michael Duff, quoted by Lisa Randall, *Knocking on Heaven's Door: How Physics and Scientific Thinking Illuminate the Universe and the Modern World*, Random House, London, 2011. p. 304.

5. In *The Trouble with Physics: The Rise of String Theory, the Fall of a Science, and What Comes Next*, published by Penguin Books, London, in 2008, Smolin wrote: 'Nearly every particle theorist with a permanent position at the prestigious Institute for Advanced Study, including the director, is a string theorist; the exception is a person hired decades ago. The same is true of the Kavli Institute for Theoretical Physics. Eight of the nine MacArthur Fellowships awarded to particle physicists since the beginning of the program in 1981 have also gone to string theorists. And in the country's top physics departments (Berkeley, Caltech, Harvard, MIT, Princeton and Stanford), twenty out of twenty-two tenured professors in particle physics who received PhDs after 1981 made their reputation in string theory or related approaches.' This quote appears in the Introduction, p. xx.

6. Lee Smolin, personal communication, 7 September 2017.

7. Sheldon L. Glashow, with Ben Bova, *Interactions: A Journey through the Mind of a Particle Physicist and the Matter of This World*, Warner Books, New York, 1988, p. 334.

8. Joseph Conlon, *Why String Theory?*, CRC Press, Boca Raton, FL, 2016, p. 226.

9. Lee Smolin, personal communication, 7 September 2017.

10. Carlo Rovelli, 'Loop Quantum Gravity: The First 25 Years', *Classical and Quantum Gravity*, **28** (2011), 153002; arXiv:gr-qc/1012.4707v5, 28 January 2012.

11. Lee Smolin, *The Trouble with Physics: The Rise of String Theory, the Fall of a Science, and What Comes Next*, Penguin Books, London, 2008, p. 284.

12. Glashow, *Interactions*, pp. 330 and 335.
13. David Gross, quoted by Carlo Rovelli, personal communication, 21 June 2017.
14. Conlon, *Why String Theory?*, p. 151.
15. Smolin, *The Trouble with Physics*, p. 275.
16. Albert Einstein, letter to the Besso family, 21 March 1955, quoted in Alice Calaprice (ed.), *The Ultimate Quotable Einstein*, Princeton University Press, Princeton, NJ, 2011, p. 113. This was a letter of condolence on the passing of his good friend Michele Besso, written less than a month before his own death.
17. John Archibald Wheeler, with Kenneth Ford, *Geons, Black Holes and Quantum Foam: A Life in Physics*, W. W. Norton, New York, 1998, pp. 148–9.
18. Fotini Markopoulou, quoted by Sally Davies, 'This Physics Pioneer Walked Away from it All', *Nautilus*, 28 July 2016. Available at http://nautil.us/issue/38/noise/this-physics-pioneer-walked-away-from-it-all.
19. Lee Smolin, personal communication, 7 September 2017.
20. Howard Burton, *First Principles: The Crazy Business of Doing Serious Science*, Key Porter Books, Toronto, 2009, p. 32.
21. Alain Connes, quoted by Carlo Rovelli, *What is Time? What is Space?*, manuscript translated by J. C. van den Berg, p. 34. Published by de Renzo Editore, Rome, 2006.
22. Roger Penrose, *The Road to Reality: A Complete Guide to the Laws of the Universe*, Vintage, London, 2005, p. 953.
23. Lee Smolin, *Three Roads to Quantum Gravity: A New Understanding of Space, Time and the Universe*, Phoenix, London, 2001, p. 211.

CHAPTER 11: GRAVITONS, HOLOGRAPHIC PHYSICS, AND WHY THINGS FALL DOWN

1. The (probably apocryphal) story goes that, following a lecture on astronomy or cosmology, a little old lady sitting at the back raised her hand and declared that this was all rubbish. Everyone knew that the world was a flat plate sitting on the back of a giant turtle (in Hindu mythology, the world is supported by an elephant, which is in turn supported by a tortoise). With an air of

smug superiority, the lecturer asked if she knew what supported the turtle. 'You're very clever, young man, very clever,' she replied. 'But it's turtles all the way down'. Various sources attribute different speakers. For example, in *A Brief History of Time*, Stephen Hawking suggests that it was the philosopher Bertrand Russell (see page 1). In *Geons, Black Holes and Quantum Foam*, John Wheeler suggests it was the philosopher William James (see page 349).

2. Lee Smolin, 'My Divergence', personal communication, 21 July 2017. The 'linking' paper is Lee Smolin, 'Linking Topological Quantum Field Theory and Nonperturbative Quantum Gravity', *Journal of Mathematical Physics*, **36** (1995), 6417; arXiv:gr-qc/9505082v2, 30 January 1996.

3. Carlo Rovelli, 'Zakopane Lectures on Loop Gravity', in John Barrett et al. (eds), *Proceedings, 3rd Quantum Geometry and Quantum Gravity School: Zakopane, Poland, February 28–March 13, 2011*; arXiv:gr-qc/1102.3660v5, 3 August 2011, p. 24.

4. Carlo Rovelli, personal communication, 31 March 2017.

5. The expression $E = mc^2$ is very familiar, but, in fact, the full expression for the relativistic energy of an object is $E^2 = p^2c^2 + m_0^2c^4$, where p is the linear momentum and m_0 is the mass of the object at rest. If it helps, we can think of this as a statement of Pythagoras' theorem. If two sides of a right-angled triangle are labelled pc and m_0c^2, then the relativistic energy is given by the square root of the hypotenuse of this triangle. For an object at rest, $p = 0$ and this expression reduces to $E_0 = m_0c^2$, where E_0 is the 'rest energy'. For an object with zero rest mass such as a photon, $m_0 = 0$ and the expression reduces to $E = pc$.

6. Lee Smolin, personal communication, 21 July 2017. The paper in question is Abhay Ashtekar, Carlo Rovelli, and Lee Smolin, 'Gravitons and Loops', *Physical Review D*, **44** (1991), 1740–55; arXiv:hep-th/9202054v1, 15 February 1992.

7. Freeman Dyson, 'The World on a String', *New York Review of Books*, 13 May 2004.

8. Tony Rothman and Stephen Boughn, 'Can Gravitons be Detected?', *Foundations of Physics*, **36** (2006), 1801–25; arXiv:gr-qc/0601043v3, 2 December 2006, p. 17.

9. Carlo Rovelli, 'Zakopane Lectures on Loop Gravity', in John Barrett *et al.* (eds), *Proceedings, 3rd Quantum Geometry and Quantum Gravity School: Zakopane, Poland, February 28–March 13, 2011*; arXiv:gr-qc/1102.3660v5, 3 August 2011, p. 15.

10. Richard P. Feynman, *QED: The Strange Theory of Light and Matter*, Penguin Books, London, 1990, p. 85.

11. Robert Oeckl, 'A "General Boundary" Formulation for Quantum Mechanics and Quantum Gravity', *Physics Letters B*, **575** (2003), 318–24; arXiv:hep-th/0306025v2, 16 October 2003, p. 1.

12. Carlo Rovelli, personal communication, 21 June 2017.

13. Carlo Rovelli, quoted in Davide Castelvecci and Valerie Jamieson, 'You Are Made of Space-time', *New Scientist*, 12 August 2006.

14. Feynman said: 'I don't like that they're not calculating anything. I don't like that they don't check their ideas. I don't like that for anything that disagrees with an experiment, they cook up an explanation—a fix-up to say "Well, it still might be true".' Quoted in P. C. W. Davies and Julian Brown (eds), *Superstrings: A Theory of Everything*, Cambridge University Press, Cambridge, UK, 1988, p. 194.

15. Eugenio Bianchi, Leonardo Modesto, Carlo Rovelli, and Simone Speziale, 'Graviton Propagator in Loop Quantum Gravity', *Classical and Quantum Gravity*, **23** (2006), 6989–7028; arXiv:gr-qc/0604044v2, 13 May 2006, p. 33.

16. The Barbero–Immirzi parameter is, in fact, already present in Ashtekar's original formulation of the constrained Hamiltonian based on his 'new variables'. It just happens to have the value i, the square root of -1. But this gives the problem that the variables of the theory are complex, requiring some care to ensure that the version of general relativity that can be recovered from the formulation is real, rather than complex. This problem can be avoided by replacing $\gamma = i$ with γ as a variable real parameter (see Rodolfo Gambini and Jorge Pullin, *A First Course in Loop Quantum Gravity*, Oxford University Press, Oxford, 2011, p. 95). The expression for the area spectrum then becomes $8\pi\gamma l_p^2 \sqrt{j(j+1)}$. Compare this with the expression given in Chapter 9, endnote 20.

CHAPTER 12: FERMIONS, EMERGENT PARTICLES, AND THE NATURE OF STUFF

1. Carlo Rovelli, personal communication, 10 April 2017.
2. Ibid.
3. Eugenio Bianchi, Muxin Han, Carlo Rovelli, Wolfgang Wieland, Elena Magliaro, and Claudio Perini, 'Spinfoam Fermions', *Classical and Quantum Gravity*, **30** (2013), 235023; arXiv:gr-qc/1012.4719v2, 21 October 2013, p. 1.
4. Wolfgang Pauli, letter to Victor Weisskopf, 17 January 1957. Quoted in Robert P. Crease and Charles C. Mann, *The Second Creation: Makers of the Revolution in Twentieth-Century Physics*, Rutgers University Press, New Brunswick, NJ, 1986, p. 209.
5. Wolfgang Pauli, letter to Victor Weisskopf, 27 January 1957. Quoted in ibid., p. 209.
6. Muxin Han and Carlo Rovelli, 'Spinfoam Fermions: PCT Symmetry, Dirac Determinant and Correlation Functions', *Classical and Quantum Gravity*, **30** (2013), 075007; arXiv:gr-qc/1101.3264v2, 6 March 2013, p. 24.
7. See Jacob Barnett and Lee Smolin, 'Fermion Doubling in Loop Quantum Gravity', arXiv:gr-qc/1507.01232v1, 5 July 2015.
8. Lee Smolin, personal communication, 8 April 2017.
9. Ibid.
10. John Archibald Wheeler, with Kenneth Ford, *Geons, Black Holes and Quantum Foam: A Life in Physics*, W. W. Norton, New York, 1998, p. 267.
11. Herbert Pfister and Markus King, 'The Gyromagnetic Factor in Electrodynamics, Quantum Theory and General Relativity', *Classical and Quantum Gravity*, **20** (2003), 205.
12. Lee Smolin, personal communication, 7 September 2017.
13. Haim Harari, 'A Schematic Model of Quarks and Leptons', *Physics Letters B*, **86** (1979), 84.
14. Sundance Bilson-Thompson, quoted in Davide Castelvecci and Valerie Jamieson, 'You Are Made of Space-time', *New Scientist*, 12 August 2006.
15. Lee Smolin, *The Trouble with Physics: The Rise of String Theory, the Fall of a Science, and What Comes Next*, Penguin Books, London, 2008, p. 254.

16. Lee Smolin, quoted in Davide Castelvecci and Valerie Jamieson, 'You Are Made of Space-time', *New Scientist*, 12 August 2006.

17. Sundance O, Bilson-Thompson, Fotini Markopoulou, and Lee Smolin, 'Quantum Gravity and the Standard Model', *Classical and Quantum Gravity*, **24** (2007), 3975–94; arXiv: hep-th/0603022v2, 21 April 2007.

CHAPTER 13: RELATIONAL QUANTUM MECHANICS AND WHY 'HERE' MIGHT ACTUALLY BE 'OVER THERE'

1. Carlo Rovelli, personal communication, 27 July 2017.

2. Lee Smolin, personal communication, 27 July 2017.

3. Carlo Rovelli, quoted by Bryan Appleyard, 'Physics Made Easy', *Sunday Times Magazine*, 11 June 2017, p. 16.

4. Carlo Rovelli, personal communication, 21 June 2017.

5. Lee Smolin, personal communication, 21 June 2017.

6. This is the 'principle of superposition'—for all linear systems, the result of combining two effects is simply the sum of these effects. The principle forms the basis of Fourier analysis (named for nineteenth-century French mathematician and physicist Joseph Fourier), in which a complex physical effect can be modeled as the linear combination of a series of simple sine waves, called a Fourier series.

7. For the avoidance of doubt, we actually take the modulus-square of these contributions, because in quantum mechanics they may be imaginary (i.e. they may contain the factor i, or $\sqrt{-1}$). For example, if the actual contribution is 'a bit $\times i$' ($i$bit), then the corresponding quantum probability is $|i\text{bit}|^2 = |(i\text{bit}) \times (-i\text{bit})| = \text{bit}^2$.

8. Carlo Rovelli, 'Space is Blue and Birds Fly Through It', arXiv:physics. hist-ph/1712.02894v2, 14 December 2017, p. 1.

9. Matteo Smerlak and Carlo Rovelli, 'Relational EPR', *Foundations of Physics*, **37** (2007), 427–45; arXiv:quant-ph/0604064v3, 4 March 2007, p. 3. The italics are mine.

10. Niels Bohr, quoted by Aage Petersen, 'The Philosophy of Niels Bohr', *Bulletin of the Atomic Scientists*, **19** (1963), 12.

11. Werner Heisenberg, *Physics and Philosophy: The Revolution in Modern Science*, Penguin, London, 1989 (first published 1958), pp. 45–6.

12. Smerlak and Rovelli, 'Relational EPR', p. 5.

13. The word 'information' has two meanings. In this description I'm using an *operational* interpretation of information, as the summary of what we *know* about a physical system. Rovelli's relational interpretation of quantum mechanics is actually based on the information content *inherent* in physical systems by virtue of the properties established by physical relationships. The two meanings are, of course, inter-related: we can't acquire operational information unless it is inherent in the systems we're studying. Rovelli explains it this way: 'I am a naturalist: I see "knowledge" as embodied in the world and not the other way around, not in terms of the world as the content of my knowledge. Of course this other foundation is viable, I just think it is less productive: I prefer to keep reminding myself that our access to the world is limited, but I accept only a worldview in which me and my knowledge are a very special case of the world and its physical correlations.' Carlo Rovelli, personal communication, 1 August 2017.

14. Albert Einstein, Boris Podolsky, and Nathan Rosen, 'Can Quantum-Mechanical Description of Physical Reality be Considered Complete?', *Physical Review*, **47**, (1935), 777–80. This paper is reproduced in John Archibald Wheeler and Wojciech Hubert Zurek (eds), *Quantum Theory and Measurement*, Princeton University Press, Princeton, NJ, 1983, pp. 138–41. This quote appears on p. 138.

15. See Andrew Whitaker, *John Stewart Bell and Twentieth-Century Physics: Vision and Integrity*, Oxford University Press, Oxford, 2016, p. 57.

16. This phrase is frequently attributed to Feynman, but it appears to have been coined by N. David Mermin, see 'Could Feynman Have Said This?', *Physics Today*, May 2004, pp. 10–11.

17. Carlo Rovelli, 'Space is Blue and Birds Fly through It', arXiv:physics.hist-ph/1712.02894v2, 14 December 2017, p. 7.

18. Albert Einstein, quoted in Maurice Solovine, *Albert Einstein: Lettres à Maurice Solovine*, Gauthier-Villars, Paris, 1956. This quote is reproduced in Arthur Fine, *The Shaky Game: Einstein, Realism and the Quantum Theory*, 2nd edition, University of Chicago Press, Chicago, 1986, p. 110.

19. Einstein, Podolsky, and Rosen, 'Can Quantum-Mechanical Description of Physical Reality be Considered Complete?', in Wheeler and Zurek (eds), *Quantum Theory and Measurement*, p. 141.

20. In a letter to Max Born dated 12 May 1952, Einstein wrote: 'Have you noticed that Bohm believes (as de Broglie did, by the way, 25 years ago) that he is able to interpret the quantum theory in deterministic terms? That way seems too cheap to me.' Einstein is referring to the de Broglie–Bohm 'pilot wave' interpretation of quantum theory, which is essentially a non-local hidden variables theory. This letter is reproduced in Max Born, *The Born–Einstein Letters*, Macmillan, New York, 1971, p. 192.

21. Roger Penrose, *The Emperor's New Mind: Concerning Computers, Minds and the Laws of Physics*, Vintage, London, 1990, p. 475.

22. Penrose wrote: 'In many standard situations of quantum measurement, the main mass displacements would occur in the *environment*, entangled with the measuring device, and in this way the conventional "environmental decoherence" viewpoint may acquire a consistent ontology.' See Roger Penrose, *Fashion, Faith and Fantasy in the New Physics of the Universe*, Princeton University Press, Princeton, 2016, p. 215.

23. John Bell, 'Against Measurement', *Physics World*, **3** (1990), 36.

24. Fotini Markopoulou and Lee Smolin, 'Quantum Theory from Quantum Gravity', *Physical Review D*, **70** (2003); arXiv:gr-qc/0311059v2, 14 June 2004.

25. Murray Gell-Mann, *Nuovo Cimento*, Suppl. 2, **4**, Series X (1958), 848–66. In a footnote on p. 859, he writes that 'any process not forbidden by a conservation law actually does take place with appreciable probability. We have made liberal and tacit use of this assumption, which is related to the state of affairs that is said to prevail in a perfect totalitarian state. Anything that is not compulsory is forbidden.' He was paraphrasing Terence Hanbury (T. H.) White, the author of *The Once and Future King*.

26. Lee Smolin, *Time Reborn: From the Crisis in Physics to the Future of the Universe*, Penguin Books, London, 2014, p. 182.

27. See Fotini Markopoulou and Lee Smolin, 'Disordered Locality in Loop Quantum Gravity States', *Classical and Quantum Gravity*, **24** (2007), 3813–24; arXiv:gr-qc/0702044v2, 21 April 2007, p. 11.

CHAPTER 14: NOT WITH A BANG: THE 'BIG BOUNCE', SUPERINFLATION, AND SPINFOAM COSMOLOGY

1. See, for example, Anna Ijjas, Paul J. Steinhardt, and Abraham Loeb, 'Pop Goes the Universe', *Scientific American*, January 2017, 32–9. This was a rather provocative article, attracting a response from no less than 33 theorists, including Alan Guth, Andrei Linde, Sean Carroll, Stephen Hawking, Lawrence Krauss, Juan Maldacena, Lisa Randall, Martin Rees, Leonard Susskind, Alexander Vilenkin, Steven Weinberg, Frank Wilczek, and Ed Witten. For a more detailed overview of the controversy, see http://www.math.columbia.edu/~woit/wordpress/?p=9289.

2. Abhay Ashtekar, 'Loop Quantum Cosmology: An Overview', *General Relativity and Gravitation*, **41** (2009), 707–41; arXiv:gr-qc /0812.0177v1, 30 November 2008, p. 1. Italics in the original.

3. Stephen W. Hawking, *A Brief History of Time: From the Big Bang to Black Holes*, Bantam Press, London, 1988, p. 51.

4. Martin Bojowald, 'Absence of a Singularity in Loop Quantum Cosmology', *Physical Review Letters*, **86** (2001), 5227–30; arXiv:gr-qc /0102069v1, 14 February 2001, p. 1. Italics in the original.

5. Ibid., p. 4. The italics are mine.

6. Carlo Rovelli, quoted in Anil Ananthaswamy, 'From Big Bang to Big Bounce', *New Scientist*, 13 December 2008, p. 32.

7. Suppose the distance between two galaxies is measured to be D at time t. If we further suppose that the distance was D_0 at some earlier reference time t_0 then the scale factor a is given by the ratio D/D_0. An expanding universe implies that D is greater than D_0 (the galaxies get further apart as time progresses) and so a is greater than 1. A contracting universe implies that D is less than D_0 (the galaxies get closer together as time progresses) and a is less than 1. Writing $D = aD_0$ and differentiating with respect to time gives: $\dfrac{dD}{dt} = D_0 \dfrac{da}{dt}$, because D_0 is a reference distance at a fixed time t_0 and so is constant with respect to time. If we now denote time derivatives using a dot notation we have $\dot{D} = \dot{a}D_0$. But we also know that $D_0 = D/a$, so we can substitute for D_0 to give: $\dot{D} = \dfrac{\dot{a}}{a}D$. Now, $\dot{D}$ is the rate of change of distance between galaxies with time, which in an expanding universe is also the *recession velocity*, v, the speed with which galaxies

are receding from each other. If we set the Hubble parameter $H = \frac{\dot{a}}{a}$, then we recover Hubble's law: $v = HD$—the speed of a distant galaxy is linearly related to its distance.

8. Abhay Ashtekar, personal communication, 11 December 2017.

9. Alejandro Corichi, personal communication, 23 November 2017.

10. The Planck density is given by the Planck mass m_p divided by the cube of the Planck length, l_p: $\rho_p = m_p / l_p^{3}$, or $2\pi c^5 / h G^2$. In fact, the critical density ρ_c in LQC is about 40 per cent of ρ_p, or about 2×10^{96} kilograms per cubic metre (see Ashtekar, 'Loop Quantum Cosmology', p. 16).

11. Ibid., p. 19.

12. See, for example, Lawrence M. Krauss, *A Universe from Nothing: Why There is Something Rather than Nothing*, Simon & Schuster, London, 2012.

13. Martin Bojowald, quoted in 'What Happened Before the Big Bang?', 1 July 2007, available at https://phys.org/news/2007-07-big.html. See also Martin Bojowald, 'Harmonic Cosmology: How Much Can We Know about a Universe before the Big Bang?', *Proceedings of the Royal Society A*, **464** (2008), 2135–50; arXiv:gr-qc/0710.4919v1, 25 October 2007, p. 14.

14. Alejandro Corichi and Parampreet Singh, 'Quantum Bounce and Cosmic Recall', *Physical Review Letters*, **100** (2008), 161302; arXiv:gr-qc/0710.4543v2, 27 March 2008, p. 1.

15. Martin Bojowald, personal communication, 10 November 2017. See also Martin Bojowald and Artur Tsobanjan, 'Effective Casimir Conditions and Group Coherent States', arXiv:math-ph/1401.5352v1, 21 January 2014.

16. Abhay Ashtekar, quoted in Anil Ananthaswamy, 'Big Bounce Cosmos Makes Inflation a Sure Thing', *New Scientist*, 13 October 2010. The italics are mine.

17. Martin Bojowald, 'Inflation from Quantum Geometry', *Physical Review Letters*, **89** (2002), 261301; arXiv:gr-qc/0206054v1, 18 June 2002.

18. See, for example, Roger Penrose, *The Emperor's New Mind: Concerning Computers, Minds and the Laws of Physics*, Vintage, London, 1990, pp. 440–7, and G. W. Gibbons and Neil Turok, 'The Measure Problem in Cosmology', *Physical Review D*, **77** (2008), 063516; arXiv:hep-th/0609095v2, 2 January 2007.

19. Abhay Ashtekar and Parampreet Singh, 'Loop Quantum Cosmology: A Status Update', *Classical and Quantum Gravity*, **28** (2011), 213001; arXiv:gr-qc/1108.0893v2, 22 August 2011, p. 77.

20. Abhay Ashtekar and David Sloan, 'Loop Quantum Cosmology and Slow Roll Inflation', *Physics Letters B*, **694** (2010), 108–12; arXiv:gr-qc/0912.4093v2, 2 October 2010, p. 1.

21. Ashtekar and Singh, 'Loop Quantum Cosmology', p. 8. The italics are mine.

22. Carlo Rovelli, personal communication, 21 June 2017.

23. See Carlo Rovelli and Francesca Vidotto, *Covariant Loop Quantum Gravity: An Elementary Introduction to Quantum Gravity and Spinfoam Theory*, Cambridge University Press, Cambridge, UK, 2014, pp. 236–7.

24. I've said 'judged by many' because there are proposals for laboratory-scale experiments to probe the Planck scale, such as those described in Igor Pikovski, Michael R. Vanner, Markus Aspelmeyer, M. S. Kim, and Caslav Bruckner, 'Probing Planck-scale Physics with Quantum Optics', *Nature Physics*, **8** (2012), 393–7.

25. The term 'angular scale' refers to the spherical functions defined on the surface of a sphere, known as *spherical harmonics*, first introduced by French physicist Pierre Simon de Laplace in 1782. Such functions are characterized by polar angles θ (a co-latitude) and Φ (essentially a longitude) and integer numbers l and m, in which l determines the number of nodes in the spherical 'standing waves' and m takes integer values from $-l$ to $+l$ (in Schrödinger's wave mechanics of the hydrogen atom, these numbers became the quantum numbers associated with the electron's orbital motion). The relation between 'angular scale'—essentially the value of θ—and l is complicated, but as a *very* rough guide we can assume $\theta = 180°/l$, i.e. angular scale decreases as l increases.

26. Ivan Agullo, Abhay Ashtekar, and William Nelson, 'The Pre-inflationary Dynamics of Loop Quantum Cosmology: Confronting Quantum Gravity with Observations', *Classical and Quantum Gravity*, **30** (2013), 085014. See also arXiv:gr-qc/1302.0254v2, 8 April 2013.

27. Abhay Ashtekar and Brajesh Gupt, 'Quantum Gravity in the Sky: Interplay Between Fundamental Theory and Observations', *Classical and Quantum Gravity*, **34** (2016), 014002. See also arXiv:gr-qc/1608.04228v2, 12 November 2016.

28. See, for example, Jakob Mielczarek, Thomas Cailleteau, Julien Grain, and Aurelien Barreau, 'Inflation in Loop Quantum Cosmology: Dynamics and Spectrum of Gravitational Waves', *Physical Review D*, **81** (2010), 104049; arXiv:gr-qc/1003.4660v2, 16 June 2010.

29. Carlo Rovelli, *Reality is Not What it Seems: The Journey to Quantum Gravity*, Allen Lane, London, 2016, p. 192.

CHAPTER 15: BLACK HOLE ENTROPY, THE INFORMATION PARADOX, AND PLANCK STARS

1. Brian Greene, *The Hidden Reality: Parallel Universes and the Deep Laws of the Cosmos*, Allen Lane, London, 2011, p. 91.

2. Shamit Kachru, Renata Kallosh, Andrei Linde, and Sandip P. Trivedi, 'de Sitter Vacua in String Theory', *Physical Review D*, **68** (2003); arXiv:hep-th/0301240v2, 10 February 2003.

3. Joe Polchinski, interview with Steve Nadis, 6 February 2006, quoted in Shing-Tung Yao and Steve Nadis, *The Shape of Inner Space: String Theory and the Geometry of the Universe's Hidden Dimensions*, Basic Books, New York, 2010, p. 234.

4. Jessie Shelton, Washington Taylor, and Brian Wecht, 'Generalized Flux Vacua', *Journal of High Energy Physics*, **02** (2007), 095; arXiv: hep-th/0607015v2, 11 August 2006.

5. Lee Smolin, *The Trouble with Physics: The Rise of String Theory, the Fall of a Science, and What Comes Next*, Penguin Books, London, 2008, p. 159.

6. L. Susskind, 'The Anthropic Landscape of String Theory', arXiv:hep-th/0302219v1, 27 February 2003. See also *The Cosmic Landscape: String Theory and the Illusion of Intelligent Design*, Little, Brown, New York, 2006.

7. Max Tegmark, *Our Mathematical Universe: My Quest for the Ultimate Nature of Reality*, Penguin Books, London, 2015; see particularly pp. 132–50.

8. Richard Feynman, in P. C. W. Davies and Julian Brown (eds), *Superstrings: A Theory of Everything?*, Cambridge University Press, Cambridge, UK, 1988, p. 194.

9. The citation reads: 'For numerous deep and groundbreaking contributions to quantum field theory, quantum gravity, string theory and geometry. With Strominger [Vafa], their joint statistical

derivation of the Bekenstein–Hawking area–entropy relation unified the laws of thermodynamics with the laws of black hole dynamics and revealed the holographic nature of quantum spacetime.' See https://breakthroughprize.org/Laureates/1/L14

10. Stephen W. Hawking, *A Brief History of Time: From the Big Bang to Black Holes*, Bantam Press, London, 1988, p. 105.

11. The Bekenstein–Hawking formula is actually $S = k_B A/4l_p^2$, where k_B is Boltzmann's constant (1.38×10^{-23} joules per kelvin), A is the surface area of the black hole, and l_p is the Planck length (1.62×10^{-35} metres).

12. Joseph Conlon, *Why String Theory?*, CRC Press, Boca Raton, FL, 2016, p. 184.

13. Ibid., p. 186.

14. Krasnov–Rovelli correspondence, 1994–6. Personal communication, from Carlo Rovelli, 7 August 2017.

15. Version 1 of Krasnov's paper was uploaded to the preprint archive on 15 March 1996, but the details (and the title) were changed in version 2, and a third version was uploaded on 27 September 1996: Kirill Krasnov, 'Counting Surface States in the Loop Quantum Gravity', *Physical Review D*, **55** (1997), 3505–13; arXiv:gr-qc/9603025v3, 27 September 1996. Rovelli's paper was uploaded to the archive on 30 March 1996: Carlo Rovelli, 'Black Hole Entropy from Loop Quantum Gravity', *Physical Review Letters*, **77** (1996), 3288; arXiv:gr-qc/9603063v1, 30 March 1996. Rovelli's derivation and Krasnov's published version 3 derivation both give $S = k_B A/16\pi l_p^2$.

16. Lee Smolin, personal communication, 21 July 2017.

17. Krasnov posted version 1 of his second paper on black hole entropy from Kiev on 21 May 1996 and the revision containing the derivation $S = (\ln 5)k_B A/8\pi\gamma\sqrt{2}l_p^2$ was uploaded from Penn State on 18 January 1997. A third version was uploaded on 20 June 1997. Kirill Krasnov, 'On Quantum Statistical Mechanics of a Schwarzschild Black Hole', *General Relativity and Gravitation*, **30** (1998), 53–68; arXiv:gr-qc/9605047v3, 20 June 1997.

18. A. Ashtekar, J. Baez, A. Corichi, and K. Krasnov, 'Quantum Geometry and Black Hole Entropy', *Physical Review Letters*, **80** (1998), 904–7; arXiv:gr-qc/9710007v1, 1 October 1997.

19. Krzysztof A. Meissner, 'Black Hole Entropy in Loop Quantum Gravity', *Classical and Quantum Gravity*, **21** (2004), 5245; arXiv:gr-

qc/0407052v1, 14 July 2004. Meissner showed that the constant γ_0 can be deduced from the relationship $\sum_{k=1}^{\infty} 2e^{-2\pi\gamma_0 \sqrt{k(k+2)/4}} = 1$. The simplest way to solve this equation is to set it up in an Excel spreadsheet and use the 'goal seek' function to set the sum equal to 1 by changing the value of γ_0. You can run the range of values of k from 1 up to any number you like (obviously, infinity is a bit impractical), but the sum converges pretty quickly and with γ_0 set to 0.23753 it doesn't change significantly for values of k greater than about 20.

20. Eugenio Bianchi, 'Entropy of Non-extremal Black Holes from Loop Gravity', arXiv:gr-qc/1204.5122v1, 23 April 2012. The approach to black hole entropy based on local horizon energy is described in Ernesto Frodden, Amit Ghosh, and Alejandro Perez, 'Quasilocal First Law for Black Hole Thermodynamics', *Physical Review D*, **87** (2013), 121503; arXiv:gr-qc/1110.4055v2, 16 December 2011.

21. Carlo Rovelli, personal communication, 21 June 2017.

22. John Archibald Wheeler, with Kenneth Ford, *Geons, Black Holes and Quantum Foam: A Life in Physics*, W. W. Norton, New York, 1998, pp. 340–1.

23. Ashtekar, et al., 'Quantum Geometry and Black Hole Entropy', p. 7.

24. Leonard Susskind, *The Black Hole War: My Battle with Stephen Hawking to Make the World Safe for Quantum Mechanics*, Little, Brown, New York, 2008, p. 294.

25. Ibid., p. 419.

26. Joseph Conlon, *Why String Theory?*, CRC Press, Boca Raton, FL, 2016, p. 119.

27. 'In summary, we see convincing reason to place AdS/CFT duality in the category of true but not proven,' Gary T. Horowitz and Joseph Polchinksi, 'Gauge/gravity Duality', in Daniele Oriti (ed.), *Approaches to Quantum Gravity: Towards a New Understanding of Space, Time and Matter*, Cambridge University Press, Cambridge, UK, 2009, p. 169; arXiv:gr-qc/0602037v3, 18 April 2006, p. 17. It's fair to say that there are a few mathematical conjectures that are broadly accepted but remain unproven. However, these refer to relationships between mathematical concepts or objects which have some established validity. In their paper, Horowitz and Polchinski mention the Riemann hypothesis, one of the great unsolved problems in

mathematics. But, without going into details, this hypothesis connects the distribution of zeros (a mathematically valid concept) in something called the Riemann zeta function (also mathematically valid). In contrast, the AdS/CFT duality is *conjectured* to be established between two theoretical structures, *neither of which has any mathematical or scientific validity.*

28. Stephen Hawking, quoted by John Baez, who was present at the 17th International Conference on General Relativity and Gravitation in Dublin; see http://math.ucr.edu/home/baez/week207.html

29. See, for example, Ahmed Almheiri, Donald Marolf, Joseph Polchinki, and James Sully, 'Black Holes: Complementarity or Firewalls?', *Journal of High Energy Physics*, **62** (2013); arXiv:hep-th/1207.3123v4, 13 April 2013.

30. Abhay Ashtekar, personal communication, 11 December 2017.

31. Carlo Rovelli and Francesca Vidotto, 'Planck Stars', *International Journal of Modern Physics D*, **23** (2014), 1442026; arXiv:gr-qc/1401.6562v4, 8 February 2014.

32. David B. Cline, Stanislaw Otwinowski, Bozena Czerny, and Agnieszka Janiuk, 'Do Very Short Gamma Ray Bursts Originate from Primordial Black Holes? Review', *International Journal of Astronomy and Astrophysics*, **1** (2011), 164–72; arXiv:astro-ph/1105.5363v1, 26 May 2011.

CHAPTER 16: CLOSE TO THE EDGE: THE REALITY OF TIME AND THE PRINCIPLES OF THE OPEN FUTURE

1. Lee Smolin, *Time Reborn: From the Crisis in Physics to the Future of the Universe*, Penguin Books, London, 2014, p. xiii.

2. 'Time', *Dark Side of the Moon*, by Pink Floyd, first released on 1 March 1973 (over *forty* years ago!).

3. Lee Smolin, personal communication, 13 July 2017.

4. Lee Smolin, 'Time Reborn', Perimeter Institute Public Lecture, 3 April 2013. Available at http://www.perimeterinstitute.ca/videos/time-reborn

5. Galileo Galilei, *Dialogues Concerning Two New Sciences*, trans. Henry Crew and Alfonso de Salvio, Macmillan, New York, 1914, pp. 249–50.

6. Smolin, *Time Reborn*, p. 30.

7. Despite Minkowski's 1908 statements, Einstein was at pains to stress that spacetime doesn't necessarily involve the 'geometrization' of

time. A spacetime metric has a *signature*: time enters with a different sign when compared with spatial coordinates. In a three-dimensional Euclidean space, if the positions are $l_1 = x_1 y_1 z_1$ and $l_2 = x_2 y_2 z_2$, the spatial interval $\Delta l = l_2 - l_1$ can be found by applying Pythagoras' theorem: $\Delta l^2 = \Delta x^2 + \Delta y^2 + \Delta z^2$. This 'distance function' is often referred to as a *metric*. It has an important property: no matter how we define the coordinate system (no matter how we define x, y, and z), the metric will always be the same (mathematicians say that it is 'invariant'). When we extend Euclidean space to include a fourth dimension of time, we must ensure that the resulting spacetime metric is likewise invariant. This means that we need a structure such as $\Delta s^2 = \Delta (ct)^2 - \Delta x^2 - \Delta y^2 - \Delta z^2$, where s is a generalized spacetime interval, t is time and c is the speed of light. This metric has the signature '$+ - - -$'. We could swop these around and define Δs^2 such that the time interval is negative and the spatial intervals are positive (with a signature '$- + + +$')—so long as these are of opposite sign Δs^2 is invariant. The choice of signs is then simply a matter of convention.

8. Smolin, *Time Reborn*, p. xvi.

9. Isaac Newton, *Mathematical Principles of Natural Philosophy*, first American edition, trans. Andrew Motte, published by Daniel Adee, New York, 1845, p. 506.

10. Stephen W. Hawking, *A Brief History of Time: From the Big Bang to Black Holes*, Bantam Press, London, 1988, p. 175.

11. Steven Weinberg, *Dreams of a Final Theory: The Search for the Fundamental Laws of Nature*, Vintage, London, 1993, p. 193.

12. Ibid., p. 200.

13. In Roberto Mangeibera Unger and Lee Smolin, *The Singular Universe and the Reality of Time*, Cambridge University Press, Cambridge, UK, 2015, p. 163, Unger writes: 'What physics has found about the workings of nature must be laboriously separated from the metaphysical pre-commitments in the light of which the significance of these findings is commonly interpreted.'

14. Eugene Wigner, 'The Unreasonable Effectiveness of Mathematics in the Natural Sciences' was the title of Wigner's Richard Courant lecture in mathematical sciences delivered at New York University on 11 May, 1959. It was published in *Communications on Pure and Applied Mathematics*, **13** (1960) 1–14.

15. Carlo Rovelli, quoted by Bryan Appleyard, 'Physics Made Easy', *Sunday Times Magazine*, 11 June 2017, p. 16.

16. Howard Burton, *First Principles: The Crazy Business of Doing Serious Science*, Key Porter Books, Toronto, 2009, p. 100.

17. Smolin, *Time Reborn*, p. 100.

18. For a good antidote to the metaphysical pre-commitment to the notion of immutable laws of nature, I strongly recommend Nancy Cartwright, *How the Laws of Physics Lie*, Oxford University Press, Oxford, 1983.

19. When the planet Uranus was discovered by William Herschel in 1781, Newton's laws of motion and the inverse-square law of gravitation could not account for its orbit. The response was not to reject the laws as false, but rather to 'save the phenomena' by hypothesizing an as-yet unobserved planet beyond Uranus. The planet Neptune was duly discovered in 1846 by the German astronomer Johann Galle, less than one degree from its predicted position. But when the same trick was attempted to explain the anomalous advance in the perihelion of Mercury—by hypothesizing an as-yet unobserved planet closer to the Sun (called Vulcan)—no such planet could be found. When confronted by potentially falsifying data, either the law itself or at least one of the auxiliary hypotheses required to apply it must be modified, but the observation or experiment cannot tell us which. In the latter case it was Newton's law of gravitation that was at fault. And so one law was replaced with another, and the game started over.

20. When faced with the puzzle of reconciling Maxwell's wave theory of electromagnetic radiation with particulate theories of matter, Einstein adopted what is generally known as a *heuristic* approach. He proposed to investigate the consequences of assuming that light 'consists of a finite number of energy quanta localized at points of space that move without dividing, and can be absorbed or generated only as complete units'. This is Einstein's light-quantum hypothesis. In fact, Einstein's 1905 paper is titled: 'On a Heuristic Point of View Concerning the Production and Transformation of Light', *Annalen der Physik*, **17** (1905), 132–48. This paper is translated and reproduced in John Stachel (ed.), *Einstein's Miraculous Year: Five Papers that Changed the Face of Physics*, Centenary edition, Princeton University Press, Princeton, NJ, 2005. The quote appears on p. 178.

21. Smolin, *Time Reborn*, p. 121. The italics are mine.
22. Anna Ijjas, Paul J. Steinhardt, and Abraham Loeb, 'Pop Goes the Universe', *Scientific American*, January 2017, 32–9.
23. Paul J. Steinhardt and Neil Turok, *Endless Universe: Beyond the Big Bang*, Weidenfeld & Nicolson, London, 2007, pp. 240, 242.
24. Smolin, *Time Reborn*, p. 126.
25. James M. Lattimer, 'The Nuclear Equation of State and Neutron Star Masses', *Annual Review of Nuclear and Particle Science*, **62** (2012), 485–515. See also arXiv:nucl-th/1305.3510v1, 15 May 2013.
26. James Lattimer, personal communication, 14 July 2017.
27. Smolin, 'Time Reborn'. Available at http://www.perimeterinstitute. ca/videos/time-reborn
28. Smolin, *Time Reborn*, p. 151.
29. Lee Smolin, 'A Real Ensemble Interpretation of Quantum Mechanics', *Foundations of Physics*, **42** (2012), 1239–61; arXiv:quant-ph/1104.2822v1, 14 April 2011.
30. Smolin, *Time Reborn*, p. 191.
31. Lee Smolin in Roberto Mangeibera Unger and Lee Smolin, *The Singular Universe and the Reality of Time*, Cambridge University Press, Cambridge, UK, 2015, p. 499.

BIBLIOGRAPHY

ASHTEKAR, ABHAY, and PULLIN, JORGE (eds), *Loop Quantum Gravity: The First 30 Years*, World Scientific, Singapore, 2017.

BAGGOTT, JIM, *The Quantum Story: A History in 40 Moments*, Oxford University Press, Oxford, 2011.

BAGGOTT, JIM, *Higgs: The Invention and Discovery of the 'God Particle'*, Oxford University Press, Oxford, 2012.

BAGGOTT, JIM, *Farewell to Reality: How Fairy-tale Physics Betrays the Search for Scientific Truth*, Constable, London, 2013.

BARBOUR, JULIAN, *The End of Time: The Next Revolution in Our Understanding of the Universe*, Phoenix, London, 2000.

BARROW, JOHN D., and TIPLER, FRANK, *The Anthropic Cosmological Principle*, Oxford University Press, Oxford, 1986.

BELL, J. S., *Speakable and Unspeakable in Quantum Mechanics*. Cambridge University Press, Cambridge, UK, 1987.

BURTON, HOWARD, *First Principles: The Crazy Business of Doing Serious Science*, Key Porter Books, Toronto, 2009.

CALAPRICE, ALICE, *The Ultimate Quotable Einstein*, Princeton University Press, Princeton, NJ, 2011.

CARROLL, SEAN, *The Particle at the End of the Universe: The Hunt for the Higgs and the Discovery of a New World*, Oneworld Publications, London, 2012.

CARTWRIGHT, NANCY, *How the Laws of Physics Lie*, Oxford University Press, Oxford, 1983.

CONLON, JOSEPH, *Why String Theory?*, CRC Press, Boca Raton, FL, 2016.

COX, BRIAN, and FORSHAW, JEFF, *Why Does E = mc^2?* Da Capo Press, Cambridge, MA, 2009.

CREASE, ROBERT P., and MANN, CHARLES C., *The Second Creation: Makers of the Revolution in Twentieth-Century Physics*, Rutgers University Press, New Brunswick, NJ, 1986.

DAVIES, P. C. W., and BROWN, J. R. (eds), *The Ghost in the Atom*, Cambridge University Press, Cambridge, UK, 1986.

DAVIES, P. C. W., and BROWN, JULIAN (eds), *Superstrings: A Theory of Everything?*, Cambridge University Press, Cambridge, UK, 1988.

DE BROGLIE, LOUIS, 'Recherches sur la Théorie des Quanta', Ph.D. Thesis, Faculty of Science, Paris University, 1924. English translation by A. F. Kracklauer.

D'ESPAGNAT, BERNARD, *Reality and the Physicist*, Cambridge University Press, Cambridge, UK, 1989.

DEWITT, BRYCE S. and GRAHAM, NEILL (eds), *The Many Worlds Interpretation of Quantum Mechanics*, Pergamon, Oxford, 1975.

DIRAC, P. A. M., *The Principles of Quantum Mechanics*, 4th edn, Oxford University Press, Oxford, 1958.

DIRAC, PAUL A. M., *Lectures on Quantum Mechanics*, Dover, New York, 2001.

EINSTEIN, ALBERT, *Relativity: The Special and the General Theory*, 100th anniversary edition, Princeton University Press, Princeton, NJ, 2015.

FEYNMAN, RICHARD, *The Character of Physical Law*, MIT Press, Cambridge, MA, 1967.

FEYNMAN, RICHARD P., *QED: The Strange Theory of Light and Matter*, Penguin, London, 1985.

FEYNMAN, RICHARD P., LEIGHTON, ROBERT B., and SANDS, MATTHEW, *The Feynman Lectures on Physics*, vol. III. Addison-Wesley, Reading, MA, 1965.

FINE, ARTHUR, *The Shaky Game: Einstein, Realism and the Quantum Theory*, 2nd edn, University of Chicago Press, Chicago, 1986.

FRENCH, A. P., *Special Relativity*, Van Nostrand Reinhold, Wokingham, 1968.

GAMBINI, RODOLFO, and PULLIN, JORGE, *A First Course in Loop Quantum Gravity*, Oxford University Press, 2011.

GARDNER, SEBASTIAN, *Kant and the Critique of Pure Reason*, Routledge, Abingdon, 1999.

GELL-MANN, MURRAY, *The Quark and the Jaguar*, Little, Brown, London, 1994.

GLASHOW, SHELDON L., with BOVA, BEN, *Interactions: A Journey through the Mind of a Particle Physicist and the Matter of This World*, Warner Books, New York, 1988.

GLEICK, JAMES, *Genius: Richard Feynman and Modern Physics*. Little, Brown, London, 1992.

GORELIK, GENNADY E., and FRENKEL, VIKTOR YA., *Matvei Petrovich Bronstein and Soviet Theoretical Physics in the Thirties*, Birkhäuser Verlag, Basel, 1994.

GREENE, BRIAN, The Elegant Universe: Superstrings, Hidden Dimensions and the Quest for the Ultimate Theory, Vintage Books, London, 2000.

GREENE, BRIAN, The Fabric of the Cosmos: Space, Time and the Texture of Reality, Allen Lane, London, 2004.

GREENE, BRIAN, The Hidden Reality: Parallel Universes and the Deep Laws of the Cosmos, Allen Lane, London, 2011.

GUTH, ALAN H., The Inflationary Universe: The Quest for a New Theory of Cosmic Origins, Vintage, London, 1998.

HAWKING, STEPHEN, A Brief History of Time: From the Big Bang to Black Holes, Bantam Press, London, 1988.

HAWKING, STEPHEN, and MLODINOW, LEONARD, The Grand Design: New Answers to the Ultimate Questions of Life, Bantam Press, London, 2010.

HEISENBERG, WERNER, Physics and Philosophy: The Revolution in Modern Science. Penguin, London, 1989 (first published 1958).

HEISENBERG, WERNER, Physics and Beyond: Memories of a Life in Science, George Allen & Unwin, London, 1971.

ISAACSON, WALTER, Einstein: His Life and Universe, Simon & Shuster, New York, 2007.

ISHAM, CHRIS J., Lectures on Quantum Theory, Imperial College Press, London, 1995.

KANT, IMMANUEL, Critique of Pure Reason, trans. J. M. D. Meiklejohn, J. M. Dent & Sons, London, 1988.

KENNEDY, J. B., Space, Time and Einstein: An Introduction, Acumen, Chesham, 2003.

KENNEDY, ROBERT E., A Student's Guide to Einstein's Major Papers, Oxford University Press, Oxford, 2012.

KRAGH, HELGE, Higher Speculations: Grand Theories and Failed Revolutions in Physics and Cosmology, Oxford University Press, Oxford, 2011

KRAUSS, LAWRENCE M., A Universe from Nothing: Why There is Something Rather than Nothing, Simon & Schuster, London, 2012.

KUHN, THOMAS S., The Structure of Scientific Revolutions, 2nd edn, University of Chicago Press, Chicago, 1970.

KUHN, THOMAS S., Black-body Theory and the Quantum Discontinuity 1894–1912, University of Chicago Press, Chicago, 1978.

LEDERMAN, LEON (with Dick Teresi), The God Particle: If the Universe is the Answer, What is the Question?, Bantam Press, London, 1993.

LEIBNIZ, GOTTFRIED WILHELM, *Philosophical Writings*, trans. Mary Morris and G. H. R. Parkinson, J. M. Dent & Sons, London, 1973.

MANGEIBERA UNGER, ROBERTO, and SMOLIN, LEE, *The Singular Universe and the Reality of Time*, Cambridge University Press, Cambridge, UK, 2015.

MEHRA, JAGDISH, *The Beat of a Different Drum: The Life and Science of Richard Feynman*, Oxford University Press, Oxford, 1994.

MOORE, WALTER, *Schrödinger: Life and Thought*, Cambridge University Press, Cambridge, UK, 1989.

MURDOCH, DUGALD, *Niels Bohr's Philosophy of Physics*, Cambridge University Press, Cambridge, UK, 1987.

NUSSBAUMER, HARRY, and BIERI, LYDIA, *Discovering the Expanding Universe*, Cambridge University Press, Cambridge, UK, 2009.

PAIS, ABRAHAM, *Subtle is the Lord: The Science and the Life of Albert Einstein*. Oxford University Press, Oxford, 1982.

PANEK, RICHARD, *The 4% Universe: Dark Matter, Dark Energy and the Race to Discover the Rest of Reality*, Oneworld, Oxford, 2011.

PENROSE, ROGER, *The Emperor's New Mind: Concerning Computers, Minds and the Laws of Physics*, Vintage, London, 1990.

PENROSE, ROGER, *The Road to Reality: A Complete Guide to the Laws of the Universe*, Vintage, London, 2005.

PENROSE, ROGER, *Fashion, Faith and Fantasy in the New Physics of the Universe*, Princeton University Press, Princeton, NJ, 2016.

RANDALL, LISA, *Warped Passages: Unravelling the Universe's Hidden Dimensions*, Penguin Books, London, 2006.

RANDALL, LISA, *Knocking on Heaven's Door: How Physics and Scientific Thinking Illuminate the Universe and the Modern World*, Random House, London, 2011.

REES, MARTIN, *Just Six Numbers: The Deep Forces that Shape the Universe*, Phoenix, London, 2000.

RINDLER, WOLFGANG, *Introduction to Special Relativity*. Oxford University Press, Oxford, 1982.

ROBINSON, ANDREW, *Einstein: A Hundred Years of Relativity*, Princeton University Press, Princeton, NJ, 2015.

ROVELLI, CARLO, *What is Time, What is Space?*, de Renzo Editore, Rome, 2006.

ROVELLI, CARLO, *Quantum Gravity*, Cambridge University Press, Cambridge, UK, 2007.

ROVELLI, CARLO, *Seven Brief Lessons on Physics*, Allen Lane, London, 2015.

ROVELLI, CARLO, *Reality is Not What it Seems: The Journey to Quantum Gravity*, Allen Lane, London, 2016.

ROVELLI, CARLO, and Vidotto, Francesca, *Covariant Loop Quantum Gravity: An Elementary Introduction to Quantum Gravity and Spinfoam Theory*, Cambridge University Press, Cambridge, UK, 2014.

SCHILPP, PAUL ARTHUR (ed.), *Albert Einstein: Philosopher-Scientist*, Library of Living Philosophers, Vol. 1, Harper & Row, New York, 1959 (first published 1949).

SCHWEBER, SILVAN S., *QED and the Men Who Made It: Dyson, Feynman, Schwinger, Tomonaga*, Princeton University Press, Princeton, NJ, 1994.

SINGH, SIMON, *Big Bang: The Most Important Scientific Discovery of All Time and Why You Need to Know About It*, Harper Perennial, London, 2005.

SMOLIN, LEE, *The Life of the Cosmos*, Oxford University Press, Oxford, 1997.

SMOLIN, LEE, *Three Roads to Quantum Gravity: A New Understanding of Space, Time and the Universe*, Phoenix, London, 2001.

SMOLIN, LEE, *The Trouble with Physics: The Rise of String Theory, the Fall of a Science, and What Comes Next*, Penguin Books, London, 2008.

SMOLIN, LEE, *Time Reborn: From the Crisis in Physics to the Future of the Universe*, Penguin Books, London, 2014.

STACHEL, JOHN (ed.), *Einstein's Miraculous Year: Five Papers That Changed the Face of Physics*, Princeton University Press, Princeton, NJ, 2005.

STEINHARDT, PAUL J., and TUROK, NEIL, *Endless Universe: Beyond the Big Bang*, Weidenfeld & Nicolson, London, 2007.

SUSSKIND, LEONARD, *The Cosmic Landscape: String Theory and the Illusion of Intelligent Design*, Little, Brown, New York, 2006.

SUSSKIND, LEONARD, *The Black Hole War: My Battle with Stephen Hawking to Make the World Safe for Quantum Mechanics*, Little, Brown, New York, 2008.

TEGMARK, MAX, *Our Mathematical Universe: My Quest for the Ultimate Nature of Reality*, Penguin Books, London, 2015.

THORNE, KIP S., *Black Holes and Time Warps: Einstein's Outrageous Legacy*, W. W. Norton, New York, 1994.

VAN FRAASSEN, BAS C., *The Scientific Image*, Oxford University Press, Oxford, 1980.

VELTMAN, MARTINUS, *Facts and Mysteries in Elementary Particle Physics*, World Scientific, London, 2003.

WEINBERG, STEVEN, *The First Three Minutes: A Modern View of the Origin of the Universe*, Basic Books, New York, 1977.

WEINBERG, STEVEN, *Dreams of a Final Theory: The Search for the Fundamental Laws of Nature*, Vintage, London, 1993.

WHEELER, JOHN ARCHIBALD, with FORD, KENNETH, *Geons, Black Holes and Quantum Foam: A Life in Physics*, W. W. Norton, New York, 1998.

WHEELER, JOHN ARCHIBALD, and ZUREK, WOJCIECH HUBERT (eds), *Quantum Theory and Measurement*, Princeton University Press, Princeton, NJ, 1983.

WHITAKER, ANDREW, *John Stewart Bell and Twentieth-Century Physics: Vision and Integrity*, Oxford University Press, Oxford, 2016.

WILCZEK, FRANK, *The Lightness of Being: Big Questions, Real Answers*, Allen Lane, London, 2009.

WOIT, PETER, *Not Even Wrong: The Failure of String Theory and the Continuing Challenge to Unify the Laws of Physics*, Vintage, London, 2007.

YAU, SHING-TUNG, and NADIS, STEVE, *The Shape of Inner Space: String Theory and the Geometry of the Universe's Hidden Dimensions*, Basic Books, New York, 2010.

ZEE, A., *Fearful Symmetry: The Search for Beauty in Modern Physics*, Princeton University Press, Princeton, NJ, 1999.

ZEE, A., *Quantum Field Theory in a Nutshell*, Princeton University Press, Princeton, NJ, 2003.

ART CREDITS

Figure 1a. *Source:* Jim Baggott, *Origins* (Oxford University Press, 2015), Figure 9. Phil Degginger / Alamy Stock Photo.

Figure 2. *Source:* Jim Baggott, *Origins* (Oxford University Press, 2015), Figure 2.

Figure 5. *Source:* Adapted from Jim Baggott, *Mass* (Oxford University Press, 2017), Figure 2.

Figure 6. *Source:* Reproduced from *American Journal of Physics*, Volume 57, Issue 2, pp. 117-20. Publication date: 02/1989. Title: Demonstration of single-electron buildup of an interference pattern. Authors: Tonomura, A.; Endo, J.; Matsuda, T.; Kawasaki, T.; Ezawa, H, with the permission of the American Association of Physics Teachers.

Figure 8. *Source:* Adapted from Jim Baggott, *Mass* (Oxford University Press, 2017), Figure 16.

Figure 9. *Source:* Adapted from Jim Baggott, *Mass* (Oxford University Press, 2017), Figure 21.

Figure 10. *Source:* Adapted from Jim Baggott, *Mass* (Oxford University Press, 2017), Figure 22, original from Cush / Wikimedia Commons / Public Domain.

Figure 11. *Source:* © 2006 by Eugene Antipov/Dual-licensed under the GFDL and CC-By-SA-2.5, 2.0, and 1.0.

Figure 13. *Source:* Adapted from Jim Baggott, *Origins* (Oxford University Press, 2015), Figure 32.

Figure 14. *Source:* Jim Baggott, *Origins* (Oxford University Press, 2015), Figure 20, original images from ESA and the Planck Collaboration; NASA / WMAP Science Team.

Figure 15. *Source* (south-pointing chariot): Joseph Needham, *Science and Civilization in China*, Volume 4: *Physics and Physical Technology*, Part II: *Mechanical Engineering* © Cambridge University Press, 1965.

Figure 16. *Source:* Lunch/ Wikimedia Commons/ CC-BY-SA-2.5

Figure 17. *Source:* Thor's hammer: Oscar Montelius, *Kulturgeschichte Schwedens von den ältesten Zeiten bis zum elften Jahrhundert nach Christus* (1906), page 309.

Figure 18. *Source:* Foxman/ Wikimedia Commons/ Public Domain.

Figure 19. *Source*: Adapted from: Fig. 2, Roger Penrose, 'Angular Momentum: An Approach to Combinatorial Space-time', first published in Ted Bastin (ed.), *Quantum Theory and Beyond* (Cambridge University Press, 1971), pp. 151–180.

Figure 20. *Source*: Adapted from: Fig. 1, Carlo Rovelli, 'Loop Quantum Gravity: The First 25 Years', *Classical and Quantum Gravity* **28** (2011), 153002.

Figure 21. *Source*: Adapted from Lee Smolin, 'Atoms of Space and Time', *Scientific American*, January 2004, p. 73.

Figure 22. *Source*: Adapted from SiBr4 - Own work, CC BY-SA 3.0, https://commons.wikimedia.org/w/index.php?curid=29799005.

Figure 23. *Source*: Adapted from Davide Castelvecci and Valerie Jamieson, 'You Are Made of Space-time', *New Scientist*, 12 August 2006; and Sundance O. Bilson-Thompson, 'A Topological Model of Composite Preons', arXiv:hep-ph/0503213v2, 27 October 2006.

Figure 25. *Source*: Republished with permission of World Scientific Publishing Co., Inc, Figure 1 from 'Loop Quantum Cosmology', by Ivan Agullo and Parampreet Singh, in Abhay Ashtekar and Jorge Pullin (Eds.), *Loop Quantum Gravity: The First 30 Years* (2017); permission conveyed through Copyright Clearance Center, Inc.

Figure 26. *Source*: Adapted from Jim Baggott, *Mass* (Oxford University Press, 2017), Figure 12.

Figure 27. *Source*: Fig. 11, Abhay Ashtekar and Brajesh Gupt, 'Quantum Gravity in the Sky: Interplay Between Fundamental Theory and Observations', *Classical and Quantum Gravity* **34** (2016), 014002 © IOP Publishing. Reproduced with permission. All rights reserved. http://iopscience.iop.org/article/10.1088/1361-6382/34/1/014002/ampdf

Figure 29. *Source*: VICTOR DE SCHWANBERG/SPL/agefotostock.

Figure 30. From Galileo Galilei, *Dialogues Concerning Two New Sciences*, translated by Henry Crew and Alfonso di Salvio, Macmillan, New York, 1914. This illustration appears on p. 249.

Photo p. 328. *Source*: Photo by Claudio Perini, taken at the Loops '07 International Conference on Quantum Gravity, 25–30 June 2007, Morelia, Mexico.

INDEX

Page numbers in italics refer to figures. Page numbers in bold type indicate entries in the glossary. n after a page number indicates a footnote or an endnote.

HIGGS

The invention and discovery of the 'God Particle'

Jim Baggott

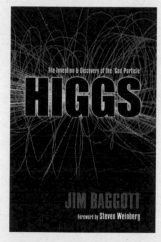

978-0-19-967957-7 | Paperback | £9.99

'A thorough and readable explanation of the lengthy hunt for the Higgs boson and why its discovery is so important.'
New Scientist

The hunt for the Higgs particle has involved the biggest, most expensive experiment ever. So exactly what is this particle? What does it tell us about the Universe? Did the discovery announced on 4 July 2012 finish the search? And was finding it really worth all the effort?

The short answer is yes. It's the strongest indicator yet that the Standard Model of physics really does reflect the basic building blocks of our Universe. Here, Jim Baggott explains the science behind the discovery, looking at how the concept of a Higgs field was invented, how the vast experiment was carried out, and its implications on our understanding of all mass in the Universe.

THE QUANTUM STORY

A history in 40 moments

Jim Baggott

978-0-19-878477-7 | Oxford Landmark
Science | Paperback | £10.99

'Jim Baggott's survey of the history of the emergence of the twentieth century's most enigmatic but successful theory is a delight to read. It is clear, accessible, engaging, informative, and thorough. It illuminates an important, revolutionary era of modern science and the varied personalities behind it.'

Peter Atkins

Almost everything we think we know about the nature of our world comes from one theory of physics. Jim Baggott presents a celebration of this wonderful yet wholly disconcerting theory, with a history told in forty episodes—significant moments of truth or turning points in the theory's development. From its birth in the porcelain furnaces used to study black body radiation in 1900, to the promise of stimulating new quantum phenomena to be revealed by CERN's Large Hadron Collider over a hundred years later, this is the extraordinary story of the quantum world.

ORIGINS

The Scientific Story of Creation

Jim Baggott

978-0-19-882600-2 | Paperback | £16.99

'The collective mind of humanity has made extraordinary progress in its quest to understand how the current richness of the physical world has emerged, and Baggott with his characteristic lucidity and erudition, has provided an enthralling account of this wonderful and still unfolding intellectual journey.'

Peter Atkins

'There are many different versions of our creation story. This book tells the version according to modern science', writes Jim Baggott. In *Origins*, he presents a unique version of the story in chronological sequence, from the Big Bang to the emergence of human consciousness 13.8 billion years later.

Cosmology, particle physics, chemistry, planetary geology, biology – it is all here, explained with clarity, in one overarching narrative. And throughout, Baggott emphasizes that the scientific story is a work-in-progress, highlighting the many puzzles and uncertainties that still remain. We have a seemingly innate desire to comprehend our own place in the Universe. Jim Baggott helps us fulfil this desire, which is driven in part by simple curiosity but also by a deeper emotional need to connect ourselves meaningfully with the world which we call home.

PLANCK

Driven by vision, broken by war

Brandon R. Brown

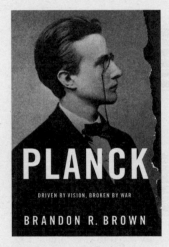

978-0-19-021947-5 | Hardback | £20.00

'An illuminating biography…Brown interweaves a gripping backstory, ranging from Planck's landmark theoretical description of blackbody radiation to his loyal advocacy for fellow physicist Lise Meitner.'

Nature

Nobel prizewinning German physicist Max Planck is credited with being the father of quantum theory, while Planck's Law was described by his close friend Albert Einstein as 'the basis of all twentieth-century physics'. But despite being one of the great scientists, Planck's personal story is not well known. Brandon R. Brown interweaves the voices and writings of Planck, his family, and his contemporaries—with many passages appearing in English for the first time—to create a portrait of a ground-breaking physicist working in Germany throughout two world wars. This gripping story of a brilliant man living through dangerous times gives Max Planck his rightful place in the history of science.

PHYSICS

A short history from quintessence to quarks

John L. Heilbron

978-0-19-874685-0 | Hardback | £10.99

'The book is effectively a short history of ideas that moves around the cultures of Europe depending on time and place, so there is a fascinating chapter on Islamic contributions.'

Network Review

'manages to pack an awful lot into that very short space...interesting and informative for non-scientists'

A Hermit's Progress

How does the physics we know today - a highly professionalised enterprise, inextricably linked to government and industry - link back to its origins as a liberal art in Ancient Greece? What is the path that leads from the old philosophy of nature and its concern with humankind's place in the universe to modern massive international projects that hunt down fundamental particles and industrial laboratories that manufacture marvels?

John Heilbron's fascinating history of physics introduces us to Islamic astronomers and mathematicians, calculating the size of the earth whilst their caliphs conquered much of it; to medieval scholar-theologians investigating light; to Galileo, Copernicus, Kepler, and Newton, measuring, and trying to explain, the universe. We visit the 'House of Wisdom' in 9th-century Baghdad; Europe's first universities; the courts of the Renaissance; the Scientific Revolution and the academies of the 18th century; the increasingly specialised world of 20th and 21st century science.